BANDPASS SIGMA DELTA MODULATORS

Bandpass Sigma Delta Modulators

Stability Analysis, Performance and Design Aspects

by

Jurgen van Engelen

Broadcom Corp., Irvine, CA, U.S.A.

and

Rudy van de Plassche

Broadcom Netherlands B.V., Bunnik, The Netherlands

KLUWER ACADEMIC PUBLISHERS

BOSTON / DORDRECHT / LONDON

A C.I.P. Catalogue record for this book is available from the Library of Congress.

ISBN 0-7923-8698-1

Published by Kluwer Academic Publishers,
P.O. Box 17, 3300 AA Dordrecht, The Netherlands.

Sold and distributed in North, Central and South America
by Kluwer Academic Publishers,
101 Philip Drive, Norwell, MA 02061, U.S.A.

In all other countries, sold and distributed
by Kluwer Academic Publishers,
P.O. Box 322, 3300 AH Dordrecht, The Netherlands.

Printed on acid-free paper

Printed in the Netherlands.

CONTENTS

LIST OF FIGURES

LIST OF TABLES

PREFACE

Sigma delta modulation has become a very useful and widely applied technique for high performance analog-to-digital (A/D) conversion of narrow band signals. Through the use of oversampling and negative feedback, the quantization errors of a coarse quantizer are suppressed in a narrow signal band in the output of the modulator. Bandpass sigma delta modulation is well suited for A/D conversion of narrow band signals modulated on a carrier, as occur in communication systems such as AM/FM receivers and mobile phones.

This book focuses on several aspects of low- *and* bandpass sigma delta modulators, in particular their stability and performance properties. The nonlinear feedback loop of the sigma delta modulator gives rise to signal dependent stability properties and deterministic effects such as dead-zones and tones which deteriorate the performance.

The first three chapters of this book serve as a general introduction to quantization, sampling and noise shaping. The effects of idle patterns and tones on the signal-to-noise ratio performance of lowpass modulators are examined more closely in chapter 4. In chapter 5, a new stability analysis method is presented, based on the describing function method and an improved quantizer model. The sampled quantizer is modeled by a linear gain *and* a phase shift. Careful analysis shows that this phase shift is in fact a phase uncertainty, and has a considerable impact on the stability analysis. The impact of the stability analysis on the loop filter design of continuous-time bandpass sigma delta modulators is analyzed in chapter 6. Finally, two design examples of continuous-time bandpass sigma delta modulators are presented in chapter 7.

Acknowledgements

We would like to thank all the people who have supported and contributed to the research and the writing of this book. In particular, we would like to thank all the people of the Electronic Circuit Design Group at Eindhoven University of Technology and the members of the Integrated Receivers Group at Philips Research Laboratories, Eindhoven, the Netherlands.

This work was supported by Nederlandse Philips Bedrijven B.V. under project number RWC-061-PS-940028-ps: "Bandpass Sigma Delta A/D Converter Architectures."

Jurgen van Engelen
August 1999

CHAPTER 1

INTRODUCTION

The increase of signal processing rates due to scaling of integrated circuit technologies has led to the replacement of analog signal processing circuits by digital signal processing systems. In audio, video, communications and many other application areas, analog techniques have been replaced by their digital counterparts. Digital signal processing has numerous advantages over analog signal processing such as flexibility, noise immunity, reliability and potential improvements in performance and power consumption by scaling of the technology. In addition, the design, synthesis, layout and testing of digital systems can be highly automated.

The advance of digital signal processing has pushed analog circuit design to the limits in more than one way. Not only are analog circuits the interface between the "analog" world and the digital signal processing system in the form of Analog-to-Digital (A/D) and Digital-to-Analog (D/A) converters, the requirements for these circuits have become increasingly higher. Analog-to-Digital conversion of signals includes two basic operations: sampling and amplitude quantization. The input signal is sampled in time and the amplitude of the input signal is mapped to a limited number of digital output codes representing the amplitude level. One way of analog-to-digital conversion is sigma delta modulation. Sigma delta modulation [1, 2] has become a very useful and widely applied technique for high-performance A/D conversion of narrow band signals. The basic thought behind sigma delta modulation is the exchange of resolution in time for resolution in amplitude. A basic diagram of a sigma delta modulator is shown in Fig. 1.1. The modulator consists of a coarse A/D converter (or ADC), a D/A converter (or DAC) and a filter placed within a feedback loop. The combination of the ADC and DAC is called a quantizer. The coarse ADC samples and quantizes the signal at its input. The number of quantization levels (or resolution) of the ADC may be as low as two, corresponding to a one-bit digital output code.

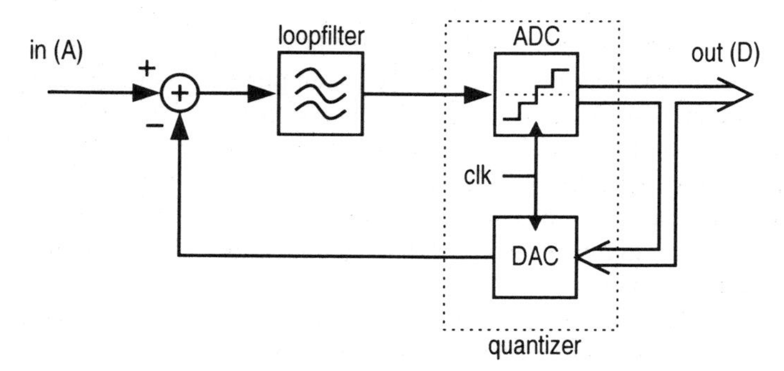

Figure 1.1: *Basic diagram of a sigma delta modulator.*

The DAC converts the resulting digital output code back to an analog signal, which is compared to the input of the modulator. The negative feedback of the loop causes the quantization errors of the coarse ADC to be suppressed for signals falling within the passband of the loop filter. The quantization errors are said to be spectrally shaped. For good suppression of the quantization errors, the sampling frequency should be much higher than the passband of the loop filter. Commonly, the passband of the filter is chosen near DC; the filter consists of integrators. In the case that the passband signal is not near DC, the modulator is called a bandpass sigma delta modulator. Bandpass sigma delta modulation is well suited for A/D conversion of signals modulated on a carrier frequency as occur in many communication systems such as AM/FM receivers and mobile phones. Even though the carrier frequency is high, the bandwidth of the signal to be converted to the digital domain is relatively small. An advantage of the sigma delta modulator over conventional A/D converters such as flash ADCs is its low sensitivity to implementation imperfections such as device mismatch. The feedback structure of the sigma delta modulator (partly) compensates for these imperfections.

Even though sigma delta modulators (SDMs) are widely used, their behavior is still not fully understood. In many ways, a sigma delta modulator resembles a nonlinear control system. The feedback loop tries to steer the ADC input signal towards zero. Consequently, the output signals of the DAC and ADC will resemble the input signal of the modulator within the passband of the loop filter. The performance of such a control system can be improved by increasing the order of the loop filter. However, the combination of a high order loop filter and the nonlinearity of the quantizer may cause instability: the internal signals of the modulator grow out of bounds or oscillate violently. When the sigma delta modulator is unstable, the output signal will no longer resemble the input signal (within the passband of the filter). This behavior is highly undesirable, and should be avoided. Unfortunately, the combination of the nonlinearity of the quantizer and the feedback loop complicates the analysis of the nonlinear (input signal dependent) behavior. Linear modeling has provided some insight in the performance and stability properties of SDMs, but does not give conclusive results. This book is aimed at the stability analysis of sigma delta modulators and the impact of the stability on the design of (bandpass) sigma delta modulators.

Chapter 2 introduces the concepts for A/D conversion: sampling in time and quantization of amplitude. The errors caused by amplitude quantization are analyzed and performance definitions relating to A/D converters are introduced. Most of these definitions can also be applied to sigma delta modulation A/D converters.

Chapter 3 gives an overview of the basic oversampling and noise shaping concepts. Several architectures for noise shaping A/D conversion are discussed and some design considerations are highlighted. One of the most straightforward architectures is the single-loop one-bit sigma delta modulator. Because this architecture can achieve a high performance and is relatively easy to design it will be the focus in the remaining chapters. Most other architectures can considered to be derivatives of the single-loop one-bit sigma delta modulator.

Chapter 4 deals with several performance issues of sigma delta modulators. A linear model for the prediction of Signal-to-Noise ratios is discussed. Deterministic effects such idle patterns and tones, and their effect on the performance of the SDM are analyzed. As the performance of an actual SDM is also determined by implementation non-idealities, this topic will also be briefly discussed.

In Chapter 5 the stability of SDMs is analyzed. Several notions and definitions concerning the stability are introduced. The stability of the SDM is analyzed by means of the describing function method. A linearized model for the sampled quantizer is developed incorporating a linear gain and a phase shift. Modeling of the phase shift of a sampled quantizer proves to be vital to explain several aspects of the stability of the SDM. Using this model, stability boundaries for loop filter parameters of lowpass modulators will be determined. A rule of thumb, derived from the analysis of lowpass modulators, is applied to bandpass modulators and stability boundaries for loop filter parameters of (a class of) bandpass SDMs are determined.

The impact of the stability properties on the design of continuous-time bandpass sigma delta modulators is discussed in Chapter 6. A simple design methodology is described and a set of continuous-time filter structures suitable for the use in bandpass SDMs is presented.

In Chapter 7, three implementations of sigma delta modulators are described. An all-digital programmable hardware set-up suitable for real-time and long-term simulation of sigma delta modulators is presented. A fourth order bandpass SDM was designed and tested to verify the feasibility of a filter structures described in the preceding chapter using a combined passive/active loop filter. The fourth order SDM was implemented using discrete components for the loop filter and served as a test case for the design of a fully integrated sixth order bandpass SDM. A prototype IC of the sixth order SDM was fully tested and its performance compared with several other reported implementations of bandpass SDMs.

The results of the research described in this book are summarized in chapter 8.

CHAPTER 2

QUANTIZATION AND SAMPLING

Real world signals are continuous in time and continuous in amplitude. In order to process these signals using digital systems, the signals have to be sampled in time and quantized to discrete amplitudes. Although in general both actions result in distortion of the original signal, the distortion resulting from sampling in time can be avoided. Quantization of the signal to discrete amplitudes always introduces errors.

2.1 Signals

Signals can be divided into numerous classes and types based on as many criteria. Two of these criteria will be considered here. A signal can be either continuous time or discrete time. Discrete time signals are defined only at certain moments in time, whereas continuous time signals are defined at every instant in time. Similarly, a signal can be either amplitude continuous or amplitude discrete. An amplitude discrete signal can only have certain amplitude values, whereas an amplitude continuous signal can have any amplitude value. In Fig. 2.1 the four resulting signal types are shown. Signals which are both amplitude continuous and continuous in time are commonly referred to as analog signals. Signals which are amplitude discrete and discrete in time are called

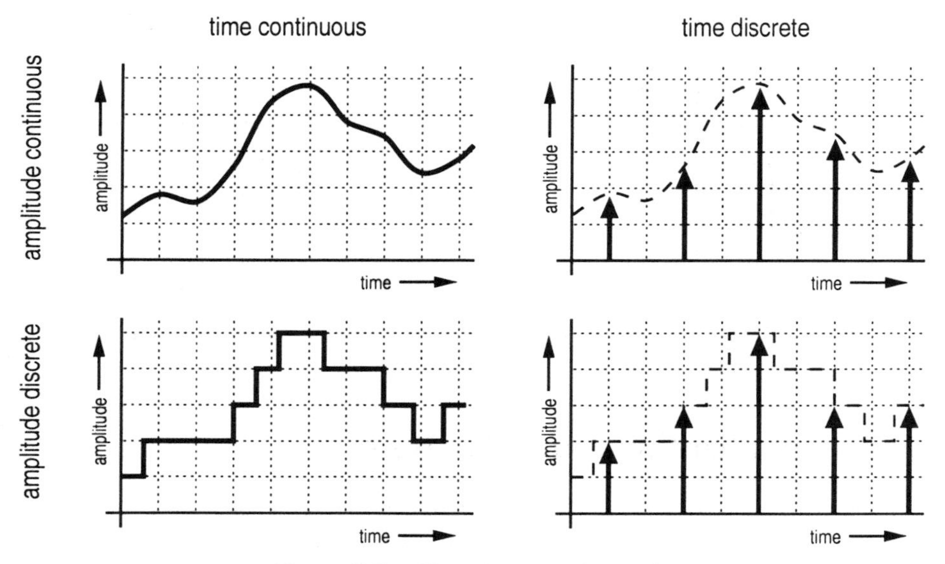

Figure 2.1: *Classification of signals.*

digital signals. Converting analog signals into digital signals requires amplitude quantization and sampling in time. These two functions are usually combined into a single Analog-to-Digital Converter (ADC), and will be discussed in detail in the following sections.

2.2 Quantization

Quantization of signals can be treated as a memoryless, time invariant and nonlinear operation. The amplitude continuous input signal is mapped on a set of discrete output values by means of rounding or a type of truncation. The input levels at which the output changes value are called threshold levels. Commonly, the input threshold levels and the output values are spaced equidistantly with step size q, resulting in a uniform quantization. In that case, the maximum absolute difference between input and output (or quantization error) equals $q/2$ (rounding). In practice the number of output values of a quantizer is limited. When B bits are used for the binary symbol identifying the active output level, a total number of 2^B output levels is possible. The maximum absolute output level then is equal to $2^{B-1}q$. In the case that the input signal exceeds the outermost quantization levels, the absolute quantization error will exceed $q/2$, and the quantizer is in *overload*. Figure 2.2 shows the input-output relations of an eight level (three bit) rounding quantizer and a four level (two bit) truncation quantizer with output offset. These quantizers are sometimes referred to as mid-tread and mid-riser quantizers respectively. For both quantizers the quantization error e_q lies within the interval $[-q/2, q/2\rangle$.

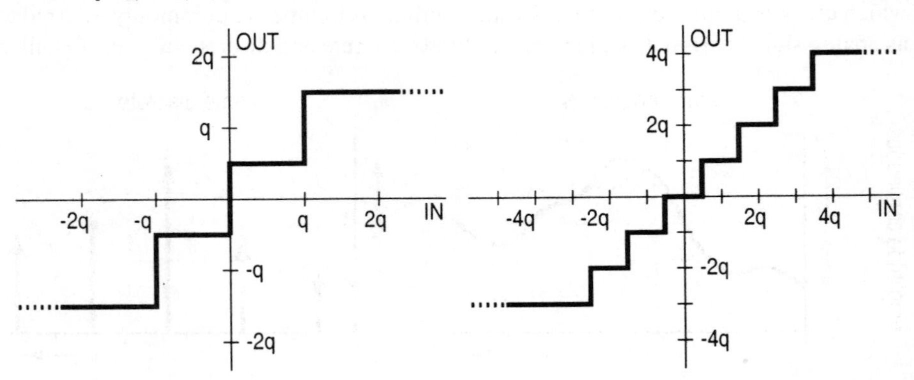

Figure 2.2: *A two bit mid-riser quantizer (left) and a three bit mid-tread quantizer (right).*

2.3 Quantization Error Analysis

Although the quantization error signal depends entirely on the input signal of the quantizer, BENNETT [3] already argued that "distortion caused by quantizing errors produces much the same sort of effects as an independent source of noise," when many

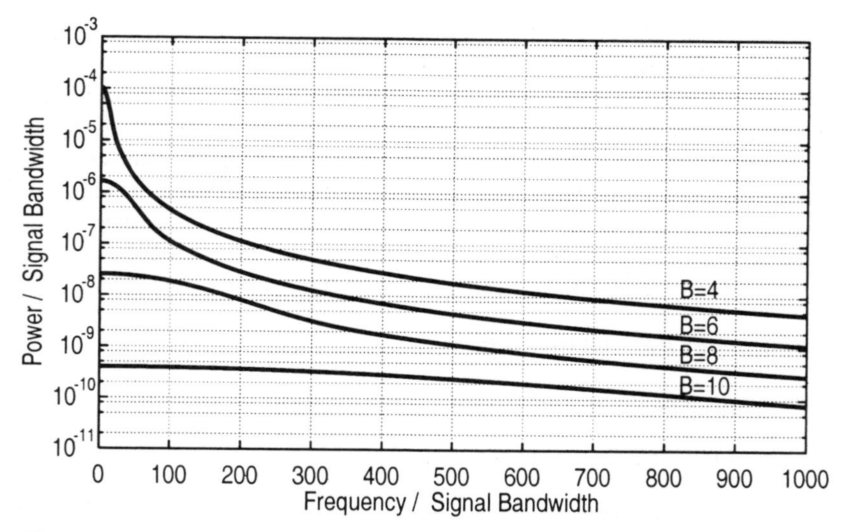

Figure 2.3: *Power spectral density of the quantization errors of a narrow band input signal quantized with B bits.*

quantization steps are used. To illustrate this, Fig. 2.3 shows the power spectral density (psd) E_q of the quantization errors of a narrow band input signal for an increasing number of quantization levels. As the number of quantization steps increases, the psd of the error signal more and more resembles the flat psd of an ideal white random signal. By calculating the autocorrelation-function of the quantization errors, BENNETT showed that for a narrow-band random input signal the power per signal bandwidth of the errors can be approximated by

$$E_q(\gamma) \approx \frac{1}{2\pi^3 4^{B-1}} \sqrt{\frac{3\kappa}{2\pi}} \sum_{n=1}^{\infty} \frac{1}{n^3} \exp\left(-\frac{3\kappa\gamma^2}{8n^2\pi^2}\right) \tag{2.1}$$

in which $\gamma = f/f_b$ is the frequency normalized to the input signal bandwidth f_b and $\kappa = q^2/\text{in}^2_{\text{rms}}$ is the ratio between the quantization step size q and the rms input level squared. For Fig. 2.3 the rms input value was chosen to be one fourth of the maximum quantizer input (no overload condition): $\kappa = 1/4^{B-3}$.

2.3.1 White Noise Approximation

The errors caused by quantization are commonly modeled by an *independent additive white noise source*, as suggested by BENNETT. By doing so, the quantization error signal is assumed to be:

- a white random (stochastic) signal;
- independent of the quantizer input signal;
- uniformly distributed within the interval $[-q/2, q/2\rangle$.

These assumptions are not valid in general. The quantization errors will never be independent of the input signal. However, WIDROW [4] showed that the errors can be uniformly distributed and uncorrelated with the input signal if certain requirements are met. By regarding quantization as a type of "area sampling" of the probability density function (pdf) of the input signal, he concluded that a modified Nyquist criterion was applicable. In the case that the Fourier transform of the pdf, called the characteristic function (cf), is band limited, the pdf of the input signal can be recovered from the quantized samples. As a result, the quantization error signal will be a uniformly distributed random signal which is uncorrelated with the input signal when

- the quantizer does not overload;
- the input signal is a random signal;
- the cf of the input signal is band limited.

A similar requirement on the (two dimensional) characteristic function of the joint-pdf of two 'samples' of the input signal is needed for the quantization errors to be white.

Less stringent requirements on the cf's of the input signal that are both sufficient and necessary were derived by SRIPAD and SNYDER [5]. Although these requirements are hardly ever met by practical signals, WIDROW showed that the results are also applicable for input signals with a Gaussian input distribution. The quantization errors will be uniformly distributed and uncorrelated with the input when the standard deviation σ of the input signal exceeds twice the quantization step size: $\sigma > 2q$. The errors will also be white if the input is a white random signal and σ is large compared to the quantization step size q.

In the case that the input signal does not satisfy these requirements, a dither signal can be used to modify the statistical properties of the input signal. The effects of dither on statistical properties of the quantization errors was investigated by LIPSHITZ et al. [6]. By adding an appropriate dither signal to the quantizer input, the quantization errors become uniformly distributed, spectrally white and statistically independent of the input signal.

Assuming that the quantization error signal is an independent additive white random signal allows straightforward calculation of some its properties. The average of the quantization error signal, which may now be called quantization noise, can be calculated from its pdf $p(e_q)$ using the expectancy operator $\mathrm{E}\{.\}$. The pdf of the uniformly distributed quantization noise is shown in Fig. 2.4. The average $\overline{e_q}$ of the

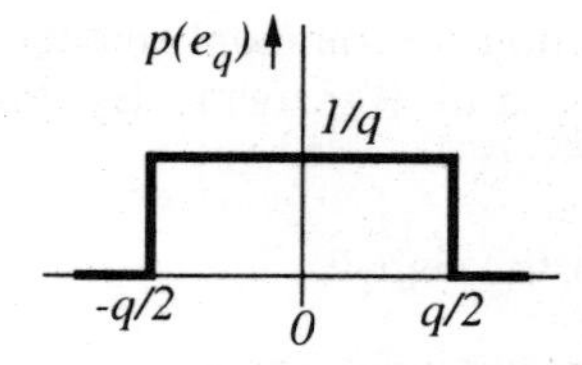

***Figure 2.4:** Probability density function (pdf) of uniformly distributed quantization noise.*

quantization noise equals:

$$\overline{e_q} = \mathrm{E}\{e_q\} = \int_{-\infty}^{\infty} e_q p(e_q)\, de_q = \int_{-q/2}^{q/2} e_q \frac{1}{q}\, de_q = 0 \tag{2.2}$$

The total power of the quantization error signal is equal to its variance and can be calculated in the same way:

$$\sigma_{e_q}^2 = \mathrm{E}\{(e_q - \overline{e_q})^2\} = \int_{-\infty}^{\infty} (e_q - 0)^2 p(e_q)\, de_q = \frac{q^2}{12} \tag{2.3}$$

2.3.2 Harmonic Distortion and Intermodulation

In the case of quantization of a non-dithered deterministic signal such as (a sum of) sinusoids, the results above do not apply. Quantization of such signals will result in harmonic distortion and intermodulation; the quantization errors will be highly correlated with the input signal. Nonlinear, signal dependent distortion is important for, for example, the field of audio signal processing. It results in audible effects, even when the system noise exceeds the distortion power.

Distortion resulting from quantizing a single sine wave has been analyzed by BLACHMAN [7]. The quantization transfer function $y(x)$ of Fig. 2.2 (left) can be considered the sum of an ideal ramp $y = x$ plus an unsymmetrical periodic sawtooth. The sawtooth can be expressed as a Fourier series. For a quantization step size $q = 1$ the quantizer transfer function can be written as

$$y(x) = x + \sum_{n=1}^{\infty} \frac{1}{n\pi} \sin(2n\pi x) \tag{2.4}$$

When the input to the quantizer is a sine wave

$$x(t) = A \sin\Theta(t) \tag{2.5}$$

the output can be expressed as a Fourier series with coefficients described in terms of Bessel functions J_p:

$$y(t) = \sum_{p=1}^{\infty} A_p \sin p\Theta(t) \qquad \text{with:} \tag{2.6}$$

$$A_p = \begin{cases} A\delta_{p1} + \sum_{n=1}^{\infty} \frac{2}{n\pi} J_p(2n\pi A) & \text{for odd } p, \\ 0 & \text{for even } p \end{cases} \tag{2.7}$$

with $\delta_{pq} = 0$ if $p \neq q$ and $\delta_{pp} = 1$. Evaluation of A_p can be done by applying a Poisson summation to (2.7), resulting in:

$$A_p = (-1)^{\frac{p-1}{2}} \frac{2}{p\pi} \sum_{k=-\lfloor A \rfloor}^{\lfloor A \rfloor} \sin\left(p \cdot \arccos(k/A)\right) \tag{2.8}$$

Quantization will not only result in an amplitude change of the fundamental (input) frequency ($p = 1$), but will also result in higher harmonics. The third order harmonic

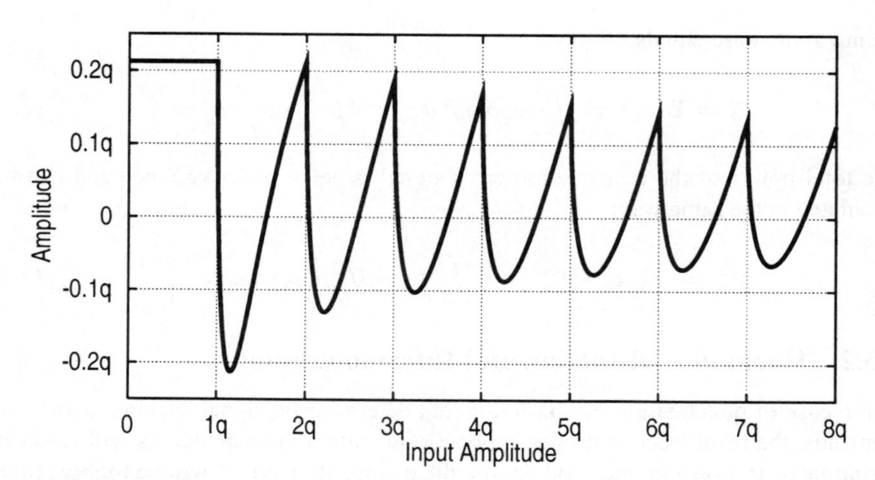

Figure 2.5: *Amplitude of the third harmonic resulting from quantizing a sine wave.*

distortion ($p = 3$) is shown in Fig. 2.5 as a function of the input amplitude. The harmonic component is almost periodic with the quantization step size and its envelope decreases slowly as the input amplitude increases. Apart from the first interval $[0, q\rangle$ the third harmonic equals zero twice in each interval $[kq, (k+1)q\rangle$.

Figure 2.6 shows the spectrum (odd harmonics) of a six bit quantized sine wave with maximum input amplitude. Similar to Fig. 2.3, most of the quantization error power is located in the lower end of the spectrum. The quantization error spectrum shows bumps and cannot considered to be white. The position of the spectral peaks

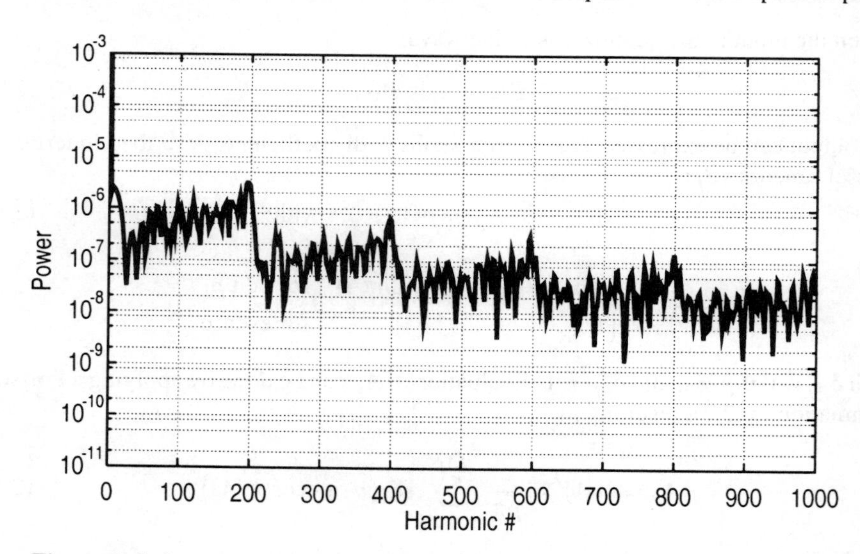

Figure 2.6: *Spectrum of a six bit quantized sine wave with maximum input amplitude.*

(bumps) was determined by CLAASEN and JONGEPIER [8], by making use of the observation that quantization results in a phase modulated signal. The power spectral density of a phase modulated signal depends on the amplitude distribution of the derivative of the modulating signal. As a result, the spectrum of a single quantized sine wave with frequency f_i and amplitude A shows peaks at frequencies which are a multiple of $f_i 2\pi A/q$:

$$f = f_i 2\pi l A/q \qquad \text{with: } l = 1,2,\ldots \tag{2.9}$$

Applying (the sum of) two sine waves with frequencies f_i and $\hat{f}_i$ and amplitudes A and $\hat{A}$ to a quantizer results in intermodulation products with frequencies $pf_i + q\hat{f}_i$. For the amplitude of these distortion components an expression similar to eq. (2.7) can be found:

$$A_{pq} = A\delta_{p1}\delta_{q0} + \hat{A}\delta_{p0}\delta_{q1} + \sum_{n=1}^{\infty} \frac{2}{n\pi} J_p(2n\pi A) J_q(2n\pi\hat{A}) \tag{2.10}$$

The distortion components close to the input frequencies f_i and $\hat{f}_i$ are often important. In communication systems, third order intermodulation products at $2f_i - \hat{f}_i$ and $2\hat{f}_i - f_i$ result in adjacent channel distortion. When the communication channels are spaced equidistantly at $\Delta f = \hat{f}_i - f_i$, the third order intermodulation products are located at $f_i - \Delta f$ and $\hat{f}_i + \Delta f$. The amplitude of these third order intermodulation distortion products is shown in Fig. 2.7 as a function of the total input amplitude $A + \hat{A}$. The amplitudes of the two input sine waves were chosen equal: $A = \hat{A}$.

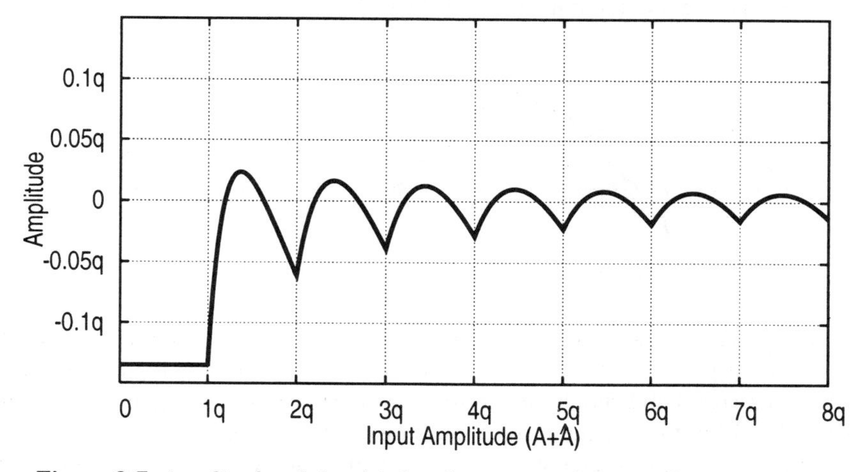

Figure 2.7: Amplitude of the third order intermodulation distortion component at $2f_i - \hat{f}_i$ as a function of the total input amplitude ($A = \hat{A}$).

2.3.3 Non-Ideal Quantization

Deviations in the positions of the output levels or input threshold levels of a quantizer results in non-ideal quantization. In the case that the deviations are small compared to the ideal step size q the distribution of the quantization errors will not change significantly and the total quantization error power will remain $q^2/12$.

However, the non-ideal quantization generally results in an increase of harmonic distortion and intermodulation. The transfer of a non-ideal quantizer can be written as the sum of an ideal ramp (x), an ideal quantization error $e_q(x)$ and the non-ideal deviations $d(x)$:

$$y(x) = x + e_q(x) + d(x) \tag{2.11}$$

The distortion caused by the non-ideal deviations can be analyzed separately. In general, the errors caused by the deviations will not counteract the distortion caused by the ideal quantizer. The deviations will result in additional harmonic distortion and intermodulation.

2.4 Sampling

Sampling of signals is a memoryless, linear operation. The continuous time input signal is sampled at discrete moments in time. Commonly, the sample moments are spaced equidistantly with interval T, resulting in uniform sampling. In the case that the sample frequency $f_s = 1/T$ exceeds twice the highest input signal frequency f_b the input signal can be fully recovered from the output samples (Nyquist Theorem). Sampling a signal at f_s samples per second causes the frequency spectrum of the input signal to be repeated at all multiples of f_s [9] (see Fig. 2.8).

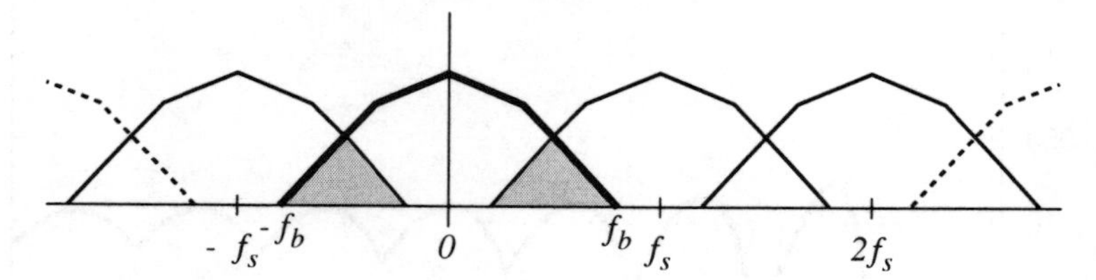

Figure 2.8: *Aliasing caused by sampling.*

Input signal components with frequencies exceeding $f_s/2$ will be folded back into the baseband of $[-f_s/2, f_s/2\rangle$ (aliasing). This type of distortion is highly correlated with the input signal and should therefore be avoided. By increasing the sample frequency or band-limiting the input signal aliasing can be prevented.

As sampling is a linear operation, the effects of sampling an amplitude quantized signal can be divided into effects on the unquantized signal and effects on the quantization error signal. In the case that the input signal satisfies the Nyquist criterion, no additional distortion will be introduced. Because the spectrum of the quantization error signal is generally not limited, it cannot satisfy the Nyquist criterion. Aliasing of the spectrum of the quantization error signal results in two effects. First, all the power of the error signal will be aliased into the baseband $[-f_s/2, f_s/2\rangle$. Secondly, in the case that the sample frequency is not much larger than twice the input signal

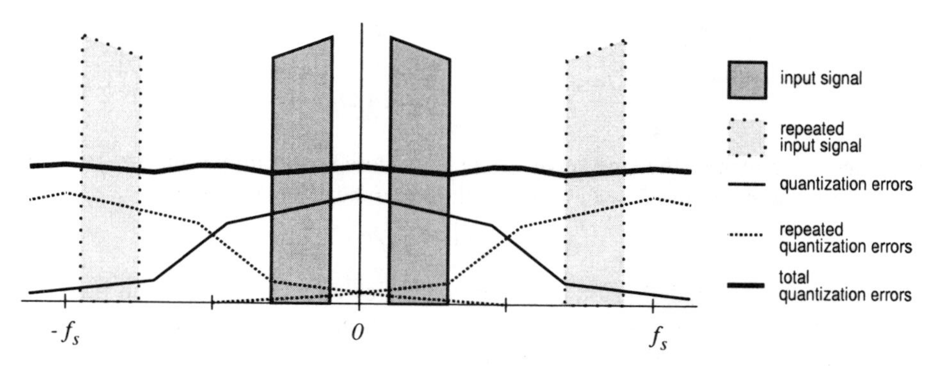

Figure 2.9: *Whitening of the quantization errors due to aliasing.*

bandwidth, aliasing will result in "whitening" of the quantization error spectrum (see Fig. 2.9). The repeated quantization error spectra at multiples of f_s will overlap the baseband and cause the overall quantization error spectrum to "flatten". When the sample frequency is very high compared to the signal frequency, the repeated spectra will not overlap significantly and the quantization errors will not appear white. The quantization errors of successive samples will become more correlated.

If the sampled quantization error signal is considered white, its psd will be constant, and the power $q^2/12$ will be uniformly distributed:

$$E_q(f) = \frac{q^2}{12f_s} \qquad \text{with: } f \in [-f_s/2, f_s/2\rangle \tag{2.12}$$

or using the angular frequency $\theta = 2\pi\frac{f}{f_s}$:

$$E_q(e^{j\theta}) = \frac{q^2}{24\pi} \qquad \text{with: } \theta \in [-\pi, \pi\rangle \tag{2.13}$$

Sampling in time and quantization of amplitude are two independent operations. The order in which these operations are performed on a signal does not affect the overall result. Therefore the quantization error analysis of sampling of an amplitude quantized continuous time signal also applies to the amplitude quantization of a discrete time signal.

2.4.1 Non-Ideal Sampling

In the case of ideal uniform sampling, the sampling moments are spaced equidistantly with interval T. Any deviation or uncertainty in the sampling interval will introduce additional distortion if the samples are treated as if they were spaced equidistantly. A deviation of Δt at a certain sample moment will result in an amplitude error ΔA. The maximum error in amplitude $\Delta A_{\max}$ will occur when the rate of change of the sampled signal is at a maximum [10]. In the case that a sine wave with amplitude A and frequency f_i is sampled, the maximum rate of change occurs around the zero crossings of the sine wave and equals

$$\left.\frac{dA\sin(2\pi f_i t)}{dt}\right|_{\max} = 2A\pi f_i \tag{2.14}$$

If the rate of change of the sine wave can be considered constant within Δt (true for $\Delta t \ll 1/f_i$), the maximum deviation in amplitude can be calculated by

$$\Delta A_{\max} = 2A\pi f_i \cdot \Delta t \tag{2.15}$$

In the case of quantized signals, this amplitude error effectively causes an increase of the quantization errors. In order to avoid a significant loss of quantizer resolution, the amplitude error should not exceed the quantization step size q. This results in a maximum value for the sampling time uncertainty

$$\Delta t < \frac{q}{2A\pi f_i} \tag{2.16}$$

This requirement is tightest when the sine wave amplitude is at a maximum. For a B bits quantizer the full scale amplitude is $A_{\max} \approx 2^B q/2$. The maximum sampling time uncertainty can be be expressed as a function of the quantizer resolution B and the input frequency f_i:

$$\Delta t_{\max} = \frac{2^{-B}}{\pi f_i} \tag{2.17}$$

For high resolution quantization of high frequency signals this results in very strict requirements for the sampling time accuracy. Random deviations in the zero crossings of a (sampling) clock signal, called *jitter*, can easily cause a reduction of the effective quantizer resolution. For a 10 MHz signal quantized with 10 bits resolution the maximum sampling time uncertainty according to (2.17) equals $\Delta t_{\max} = 31$ ps.

2.5 Performance Definitions

Although sampling and quantization are two separate signal operations, they are usually combined to form an Analog-to-Digital Converter (ADC). The performance of such an ADC can be described by several specifications [10]. A number of these parameters concern the effects of quantization errors. One of the most important specifications is the Signal-to-Noise Ratio (SNR), which is measured at the output of the ADC. The "noise" consists of system noise such as thermal and $1/f$-noise, and quantization errors. Assuming that the system noise is well below the quantization errors, the maximum SNR of an ideal B-bits ADC can be determined.

The maximum amplitude of a sine wave which does not cause the quantizer to overload equals $A_{\max} = 2^B q/2$. The root mean square (rms) amplitude of this sine wave equals $A_{\max,\mathrm{rms}} = 2^B q/2\sqrt{2}$. The total quantization error power in the Nyquist frequency range $[-f_s/2, f_s/2\rangle$ is $q^2/12$, which gives an rms amplitude of $q/\sqrt{12}$. The maximum SNR of an ideal B bits converter equals

$$\mathrm{SNR}_{\max} = 20\log_{10}\left(\frac{2^B q/2\sqrt{2}}{q/\sqrt{12}}\right) = B \times 6.02 + 1.76 \text{ dB} \tag{2.18}$$

Using this result, the maximum SNR of an ADC measured within a certain bandwidth can be expressed as an Effective Number Of Bits (ENOB):

$$\mathrm{ENOB} = \frac{\mathrm{SNR}_{\max} - 1.76}{6.02} \tag{2.19}$$

The expression for the maximum SNR in (2.18) was determined using the maximum sine wave amplitude which does not cause the quantizer to overload. In reality, this amplitude cannot be obtained, and the actual maximum sine wave amplitude lies between:

$$(2^B - 1)q/2 \leq A_{\max} \leq 2^B q/2 \tag{2.20}$$

Introducing the correction factor S_c, the maximum amplitude can be written as

$$A_{\max} = (2^B - S_c)q/2 \qquad \text{with:} \qquad 0 \leq S_c \leq 1 \tag{2.21}$$

In the case of a one-bit quantizer, this correction factor can be calculated [10]. Applying a sine wave to the input of a one-bit quantizer will result in a square wave output. The fundamental of the corresponding series expansion has a maximum amplitude of $A_{\max} = \frac{4}{\pi}q/2$. Using (2.21), the correction factor for a one-bit quantizer can be determined:

$$S_c = 2 - \frac{4}{\pi} \tag{2.22}$$

This value can be used as an estimate for the correction factor of all quantizers. The actual maximum sine wave amplitude of a B bits quantizer can be approximated by:

$$A_{\max} = (2^B - 2 + \frac{4}{\pi})q/2 \tag{2.23}$$

As a result, the expression (2.18) for the maximum SNR of a B bits quantizer changes into:

$$\mathrm{SNR}_{\max} = 20\log_{10}(2^B - 2 + \frac{4}{\pi}) + 1.76 \text{ dB} \tag{2.24}$$

Applying a sine wave to the input of the ADC will cause harmonic distortion components of the input signal to appear in the output. By including the total harmonic distortion (THD) in the calculation of the noise component, the Signal-to-Noise-and-Distortion Ratio (SNDR) is determined.

Another related specification is the Dynamic Range (DR). The definition of this specification in literature is ambiguous. The following definition will be used here:

The dynamic range (DR) of a system equals the ratio between the maximum possible signal amplitude and the minimum detectable signal amplitude within a certain bandwidth at the output of a system.

A signal component is considered detectable when its power exceeds the system noise power within a certain bandwidth. In many cases, the minimum detectable signal equals the idle channel noise of a system. The maximum possible amplitude of a signal depends on the type of signal. In the most common case of a sine wave, the maximum amplitude is equal to the maximum amplitude of the fundamental harmonic.

Important specifications concerning the linearity of a system are the Spurious Free Dynamic Range (SFDR), intermodulation products (IM) and intermodulation intercept points (IP). The SFDR of a system is defined as the ratio between the maximum signal component (usually a single sine wave) and the maximum distortion component. The definition of the SFDR is shown in Fig. 2.10.

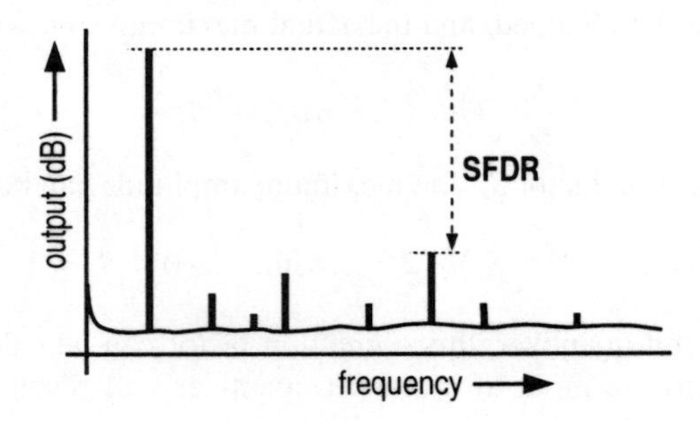

Figure 2.10: Definition of SFDR.

The third order intermodulation product (IM3) is determined using two input sine waves with frequencies f_1 and f_2. The IM3 is defined as the ratio between the carrier input power and the power of the distortion component at frequency $2f_1 - f_2$ or $2f_2 - f_1$ (see Fig. 2.11). The IP3 is defined as the input power at which the third order

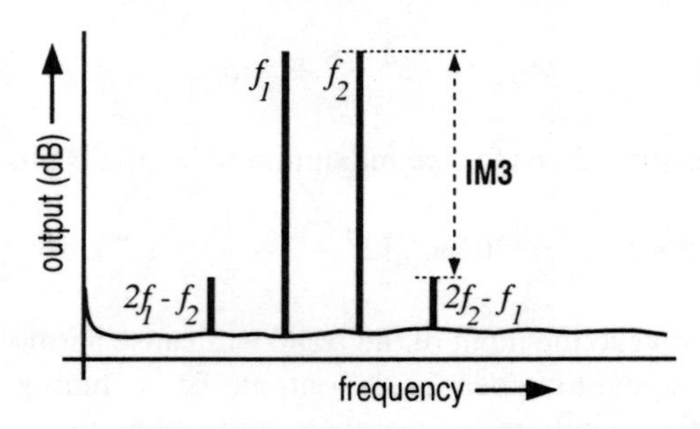

Figure 2.11: Definition of IM3.

intermodulation distortion is equal to the input power. For systems with a large linear operating range, the third order intermodulation intercept point (IP3) can be related to the IM3. For such systems, the most common nonlinearity is the loss of gain at high input amplitudes. The transfer function of the system can be approximated by

$$y(x) = g \cdot x + h \cdot x^3 \tag{2.25}$$

in which g and h are constants. For systems with a large linear range, the constant h will be very small compared to g. The third order term in (2.25) results in harmonic distortion and intermodulation products. Because of the third order, the intermodulation distortion will increase three times as fast (on log scale) as the linear term with increasing an input signal (see Fig. 2.12). The IP3 can be calculated from the IM3 using

$$\mathrm{IP3} = \mathrm{PX} - \frac{\mathrm{IM3}}{2}\ \mathrm{(dB)} \tag{2.26}$$

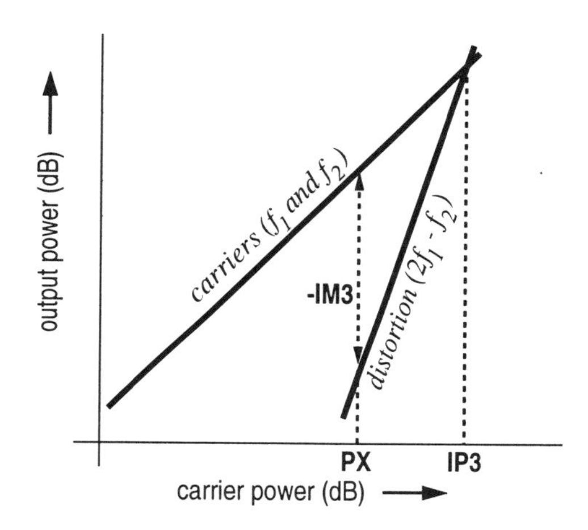

Figure 2.12: *Relationship between IM3 and IP3.*

Note that equation (2.25) does not fit the behavior of ADCs very well. If the input signal to an ADC is decreased, less quantization steps are used and the distortion relative to the input amplitude increases. In effect, the parameter h in (2.25) increases when the input amplitude decreases.

Definitions of more suitable linearity specifications for ADCs such as integral linearity (INL) and differential linearity (DNL) can be found in [10]. These definitions are intended for ADCs sampled at the Nyquist rate, and cannot readily be applied to oversampled noise shaping ADCs which will be described in the next chapter.

2.6 Conclusions

In this chapter the effects of quantization and sampling are discussed. The quantization errors are analyzed. The requirements for which the quantization errors can assumed to be white and uniformly distributed are presented. Additionally, a number of performance definitions for ADCs have been introduced. Most of these definitions can be applied directly in the context of oversampled noise shaping ADCs.

CHAPTER 3

NOISE SHAPING CONCEPTS

Noise shaping is a technique for increasing the resolution of analog-to-digital converters and digital-to-analog converters. The basic thought behind noise shaping is the exchange of resolution in time for resolution in amplitude. By increasing the resolution in time (oversampling) and by applying error feedback, the resolution and the theoretically achievable SNR of an ADC or DAC can be increased.

3.1 Oversampling

In the case that the sample frequency f_s of an ADC is equal to twice the maximum input signal frequency f_b (Nyquist), all the quantization noise power is located inside the signal bandwidth $[-f_b, f_b\rangle$. When the sample frequency f_s exceeds twice the maximum input signal frequency f_b, the ADC is said to be *oversampled*; the ratio $f_s/2f_b$ is called the oversampling ratio (OSR). When the quantization errors are assumed to be uniformly distributed within $[-f_s/2, f_s/2\rangle$, the quantization noise inside the signal bandwidth will decrease when the oversampling ratio is increased. Figure 3.1 shows the power spectral density of the quantization noise for an ADC sampled at Nyquist rate (OSR = 1) and for an ADC oversampled with a factor of four (OSR = 4). Although

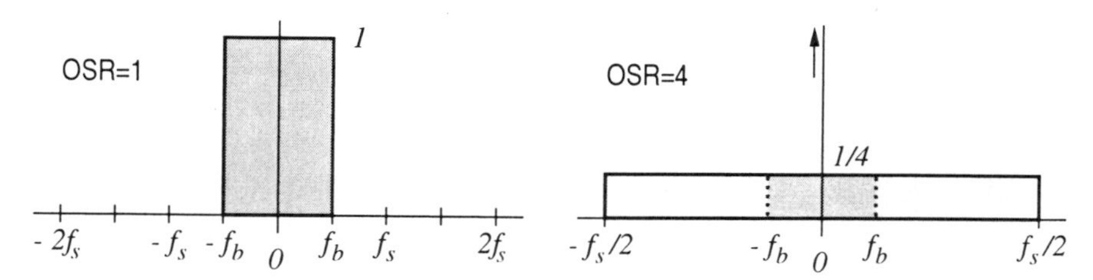

Figure 3.1: *Power spectral density of the quantization noise for an ADC sampled at Nyquist rate (OSR = 1) and an ADC oversampled with a factor of four (OSR = 4).*

the total power (area) of the quantization noise is the same, the amount of quantization noise that falls within the signal band is substantially lower when the ADC is oversampled. The rms amplitude of the noise inside the signal band can be calculated using eq. (2.12):

$$e_{q,\mathrm{rms}}^2 = \int_{-f_b}^{f_b} E_q(f)\, df = \int_{-f_b}^{f_b} \frac{q^2}{12 f_s}\, df = \frac{q^2}{12} \cdot \frac{2f_b}{f_s} = \frac{q^2}{12} \cdot \frac{1}{\mathrm{OSR}} \tag{3.1}$$

The maximum SNR inside the signal bandwidth of an ideal oversampled ADC equals

$$\mathrm{SNR}_{\mathrm{max}} = B \times 6.02 + 1.76 + 10\log_{10}(\mathrm{OSR}) \text{ dB} \quad (3.2)$$

Compared to an ADC sampled at Nyquist rate (2.18), the maximum SNR has increased with $10\log_{10}(\mathrm{OSR})$. By oversampling an ADC with a factor of four, its ENOB can be increased by one bit. The increased resolution in time can be used to increase the resolution in amplitude. Note that this increase relies on the quantization errors being white. If a low resolution quantizer (B is small) is highly oversampled, the quantization errors will not be whitened by aliasing. Successive quantization errors will become highly correlated and the oversampling will not result in an increase of the SNR (see sec. 2.4).

The definition of oversampling ratio given above is based on the fact that the input signal is a baseband signal, i.e. its frequency range is $[0, f_b]$. However, oversampling can also be used when the input signal is a bandpass signal. A more general definition of oversampling ratio would therefore be:

The oversampling ratio (OSR) **is defined as the ratio between the sample frequency f_s and twice the signal bandwidth** BW.

For a baseband signal the bandwidth is defined as the maximum signal frequency $\mathrm{BW} = f_b$. The bandwidth of a bandpass signal is defined as the difference between the maximum and minimum signal frequency $\mathrm{BW} = f_{\mathrm{max}} - f_{\mathrm{min}}$.

3.2 Error Feedback

The use of feedback to reduce errors and control the state of a (nonlinear) system is well known from control theory. In 1946 DELORAINE et al. patented such a feedback system using impulses for signal transmission [11]. DE JAGER [12] used a similar technique for analog-to-digital conversion. The output of a sampled coarse quantizer was filtered (using an integrator) and subtracted from the input signal in order to reduce the quantization errors. A more effective method to reduce the quantization errors was described by CUTLER [13] and SPANG and SCHULTHEISS [14]. Instead of feeding back the output signal, the quantization error signal is filtered and subtracted from the input signal. The resulting error feedback coder is shown in Fig. 3.2. The underlying concept of this error feedback system is to predict and correct for the next quantization error using previous quantization errors. The output (o) of the error feedback system can be written in terms of the input (i) and the quantization error (e_q). Without loss of generality, these signals can considered to be discrete time signals and the loop filter H can be considered a discrete time filter [15]. Using Fig. 3.2, the z-transform $O(z)$ of the quantizer output signal can be written as

$$O(z) = E_q(z) + X(z) = I(z) + (1 - H(z)) \cdot E_q(z) \quad (3.3)$$

in which z is the z-transform variable and the capitalized symbols indicate the z-transform of the corresponding signals. The output consists of the input signal and a

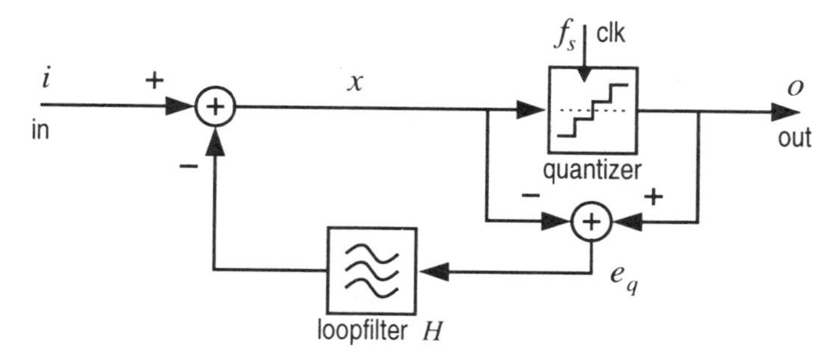

Figure 3.2: Error feedback coder or noise shaper.

filtered version of the quantization errors. The quantization errors are filtered with $1 - H(z)$, and will be eliminated from the output for $H(z) = 1$. As the quantization "noise" is spectrally shaped with $1 - H(z)$, the system of Fig. 3.2 is also referred to as a noise shaper.

It should be noted that the quantization error signal e_q is not an independent signal, but depends entirely on the input signal i, the transfer of the quantizer and the loop filter H. Nonetheless, the analysis given above applies unconditionally as no assumptions were made concerning the nature of the quantization error signal e_q.

The transfer from the quantization errors to the output of the noise shaper is commonly referred to as the Noise Transfer Function (NTF); the transfer from the input to the output of the noise shaper is called the Signal Transfer Function (STF). The output of the noise shaper can be written as

$$O(z) = \mathrm{STF}(z) \cdot I(z) + \mathrm{NTF}(z) \cdot E_q(z) \tag{3.4}$$

with:

$$\mathrm{NTF}(z) = 1 - H(z) \tag{3.5}$$
$$\mathrm{STF}(z) = 1 \tag{3.6}$$

The design of the NTF (and the loop filter) depends on the type of signals which have to be sampled and quantized. The loop filter acts as a prediction filter for the next quantization error. For slowly varying input signals, the most simple prediction for the next quantization error is the preceding quantization error. In that case, the loop filter is a delay: $H(z) = z^{-1}$, resulting in a first order noise shaper. The amplitude response $|\mathrm{NTF}(e^{j\theta})|^2$ of the noise transfer function equals

$$|\mathrm{NTF}(e^{j\theta})|^2 = |1 - e^{-j\theta}|^2 = 2 - 2\cos(\theta) \tag{3.7}$$

and is shown in Fig. 3.3. The angular frequency θ is related to the z-transform variable by:

$$z = r \cdot e^{j\theta} \qquad \text{with:} \quad \theta = 2\pi\frac{f}{f_s} \tag{3.8}$$

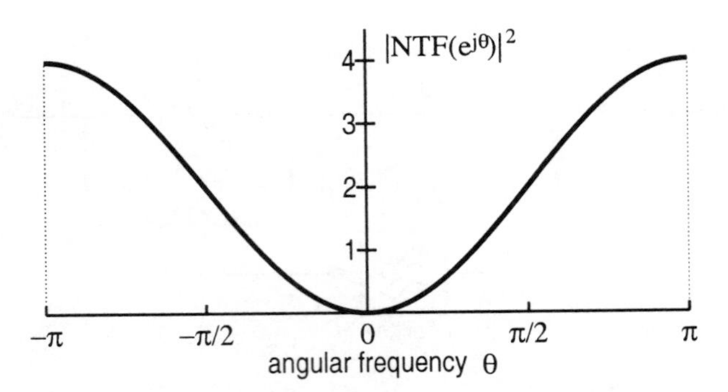

Figure 3.3: *Amplitude of the NTF of a noise shaper with* $H(z) = z^{-1}$

For low frequencies ($\theta \approx 0$) the amplitude response $|\mathrm{NTF}(z)| \approx 0$, and the quantization errors will be suppressed in the output of the noise shaper. For high frequencies ($\theta > \pi/3$) the quantization errors are amplified.

As this noise shaper suppresses the quantization errors at low frequencies it is commonly referred to as a lowpass noise shaper. Similarly, the loop filter can be designed to suppress the quantization error in a pass band, resulting in a *bandpass* noise shaper [16].

The prediction of the quantization errors can be improved by using more information, i.e. more preceding samples, of the quantization error signal. Using more samples increases the order of loop filter and the order of the noise shaper. Unfortunately, the order of the loop filter cannot be increased without limits. The nonlinearity of the quantizer within the feedback loop of the noise shaper causes stability problems if the order of the loop filter is higher than two. The stability problems are most profound for very low resolution quantizers, which exhibit a significant amount of nonlinear behavior.

3.3 Architectures

The implementation of the error feedback coder or noise shaper suffers from practical problems. Together with the stability issue for high order loop filters, the realization of the analog subtracters within the feedback path constitutes one of the major problems. The suppression of the quantization errors strongly depends on the accuracy of the subtracter that results in the quantization error signal e_q. If the transfer from the quantizer input x to the quantization error signal e_q deviates from the ideal value of unity, the quantization noise suppression is reduced significantly.

3.3.1 Sigma Delta Modulator

In order to avoid the implementation of the analog subtracter, INOSE, YASUDA and MURAKAMI [17] suggested to move the loop filter from the feedback path to the feedforward path and use the output signal as the feedback signal. This resulted in the

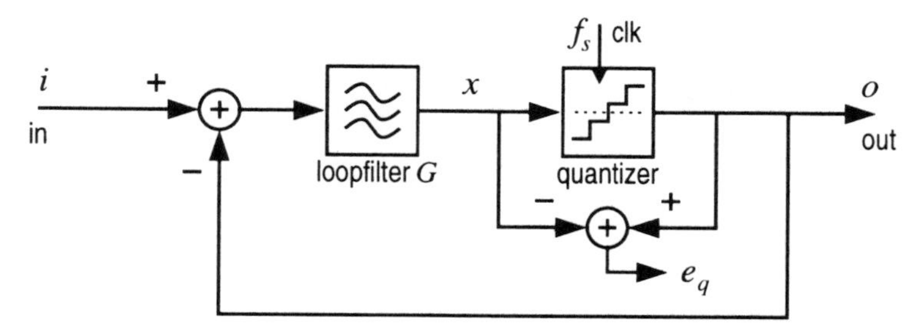

Figure 3.4: *Sigma Delta Modulator (SDM).*

Sigma Delta Modulator (SDM) shown in Fig. 3.4. The output signal of the SDM can again be written in terms of the quantization error $E_q(z)$ and the input signal $I(z)$:

$$O(z) = \frac{G(z)}{1+G(z)} \cdot I(z) + \frac{1}{1+G(z)} \cdot E_q(z) \tag{3.9}$$

The SDM of Fig. 3.4 has the following NTF and STF:

$$\mathrm{NTF}(z) = \frac{1}{1+G(z)} \tag{3.10}$$

$$\mathrm{STF}(z) = \frac{G(z)}{1+G(z)} \tag{3.11}$$

The quantization errors are filtered with $1/(1+G(z))$ and will be suppressed in the output of the SDM when $|G(z)| \gg 1$. The input signal is also filtered, but for $|G(z)| \gg 1$, the STF will be approximately unity. In order to have the same NTF as the noise shaper from Fig. 3.2, the SDM's loop filter $G(z)$ should satisfy

$$\frac{1}{1+G(z)} = 1 - H(z) \tag{3.12}$$

resulting in the relationship between the SDM loop filter $G(z)$ and the noise shaper loop filter $H(z)$:

$$G(z) = \frac{H(z)}{1-H(z)} \tag{3.13}$$

For the simplest case of $H(z) = z^{-1}$, the SDM loop filter becomes an integrator, which effectively is a lowpass filter. The large gain of the integrator for low frequencies ($z \approx 1$) results in the suppression of the quantization errors for baseband signals. The resulting STF and NTF are

$$\mathrm{NTF}(z) = 1 - z^{-1} \tag{3.14}$$

$$\mathrm{STF}(z) = z^{-1} \tag{3.15}$$

The STF shows that the input signal is delayed by one sample period. Ideally, a delay does not introduce amplitude distortion and has a linear phase characteristic.

3.3.2 Multi Stage Noise Shaping (MASH)

A solution to the stability problems of a noise shaper with a high order loop filter was suggested by HAYASHI et al. [18]. They proposed the use of several stages instead of a single high order loop filter to reduce the quantization errors of a coarse quantizer. A basic diagram of the so-called MASH [19] structure with two stages is shown in Fig. 3.5. The quantization error signal e_{q1} of the first SDM (top) is used as the input of

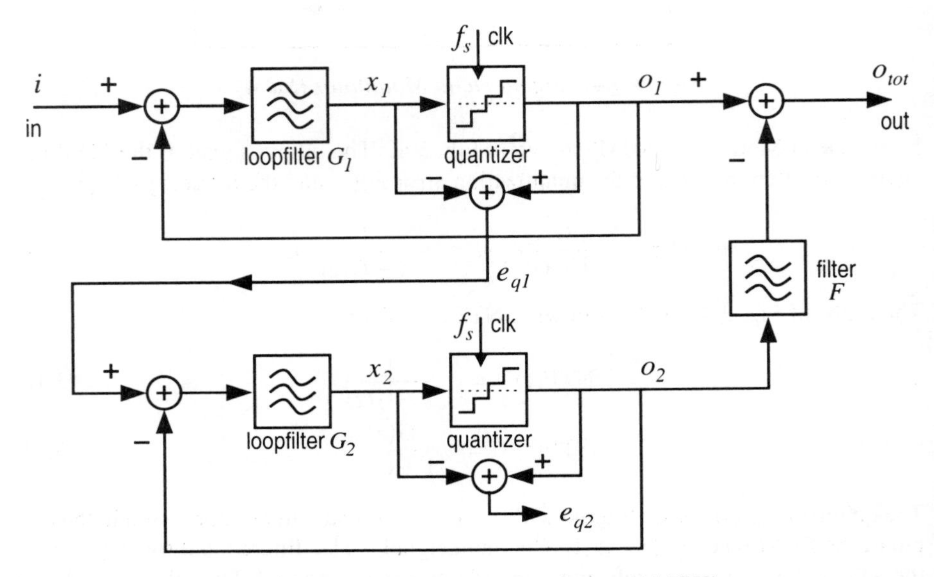

Figure 3.5: *Multi stage noise shaper (MASH).*

a second modulator. The output o_2 of this modulator thus contains a quantized estimate of the quantization error signal e_{q1}. The output of the second modulator is filtered and combined with the output of the first SDM to eliminate the quantization errors of the first modulator. The output of the first modulator equals

$$O_1(z) = \frac{G_1(z)}{1+G_1(z)} \cdot I(z) + \frac{1}{1+G_1(z)} \cdot E_{q1}(z) \tag{3.16}$$

and the output of the second SDM can be written as

$$O_2(z) = \frac{G_2(z)}{1+G_2(z)} \cdot E_{q1}(z) + \frac{1}{1+G_2(z)} \cdot E_{q2}(z) \tag{3.17}$$

The combined output O_{tot} of the two modulators can be expressed as

$$O_{tot}(z) = \frac{G_1(z)}{1+G_1(z)} \cdot I(z) + \left(\frac{1}{1+G_1(z)} - \frac{F(z)G_2(z)}{1+G_2(z)}\right) \cdot E_{q1}(z) - \frac{F(z)}{1+G_2(z)} \cdot E_{q2}(z) \tag{3.18}$$

Clearly, the quantization errors of the first SDM will be eliminated if

$$\frac{1}{1+G_1(z)} = \frac{F(z)G_2(z)}{1+G_2(z)} \tag{3.19}$$

The remaining quantization errors of the second modulator are now filtered with

$$\mathrm{NTF}_2(z) = -\frac{F(z)}{1+G_2(z)} = -\frac{1}{1+G_1(z)} \cdot \frac{1}{G_2(z)} \tag{3.20}$$

which can be of the second order even if both $F(z)$ and $G_2(z)$ are of the first order. In the case that two first order SDMs are used, the loop filters $G_1(z)$ and $G_2(z)$ are integrators:

$$G_1(z) = G_2(z) = \frac{z^{-1}}{1-z^{-1}} \tag{3.21}$$

According to (3.19), in order to eliminate the quantization errors of the first SDM, the equalization filter $F(z)$ should be

$$F(z) = \frac{1-z^{-1}}{z^{-1}} \tag{3.22}$$

which is a non causal filter. This problem can be solved by delaying the output o_1 of the first SDM by one sample period. The equalization filter $F(z)$ then changes into

$$F(z) = 1-z^{-1} \tag{3.23}$$

The remaining quantization error e_{q2} of the second SDM in the output of the MASH structure is now filtered or "shaped" by a second order function, while the input signal is delayed by two sample periods:

$$\mathrm{NTF}_2(z) = (1-z^{-1})^2 \tag{3.24}$$

$$\mathrm{STF}(z) = z^{-2} \tag{3.25}$$

The quantization error suppression has been improved without increasing the order of the loop filter of the SDM. In the case of an analog implementation, the MASH structure is sensitive to mismatch. For example, any deviation in the transfer of the filters $G_1(z)$, $G_2(z)$ and $F(z)$ will cause a mismatch in the elimination of the quantization error of the first SDM. If eq. (3.19) is no longer satisfied, the output of the MASH structure will also contain a fraction of the quantization error of the first SDM. This fraction of the quantization error e_{q1} is suppressed using a single loop only, and will cause a significant degradation in the quantization error suppression.

The MASH structure of Fig. 3.5 can be extended with additional stages in order to increase the effective order of the NTF and thereby improving the quantization error suppression. Because each low order stage operates independently, adding additional stages does not cause stability problems.

3.3.3 Other Architectures

Most realizations of oversampled noise shapers have one of the aforementioned structures. However, alternative, more exotic structures have been proposed to avoid stability problems or, for example, reduce the oversampling ratio. Some of these structures will be discussed briefly.

A. Parallel Sigma Delta Modulators

As was shown in sec. 3.2, a noise shaper or sigma delta modulator suppresses the quantization errors only in a part of the frequency spectrum. For other frequencies, the quantization errors are amplified (see Fig. 3.3). In order to increase bandwidth in which the quantization errors will be sufficiently attenuated, several architectures have been proposed placing multiple noise shapers or sigma delta modulators in parallel. An intuitive approach is to divide the required bandwidth into several smaller frequency ranges (or channels) and use a separate SDM for each channel [20]. Figure 3.6 shows a block-diagram of such a multi-band parallel SDM. The input is separated into K chan-

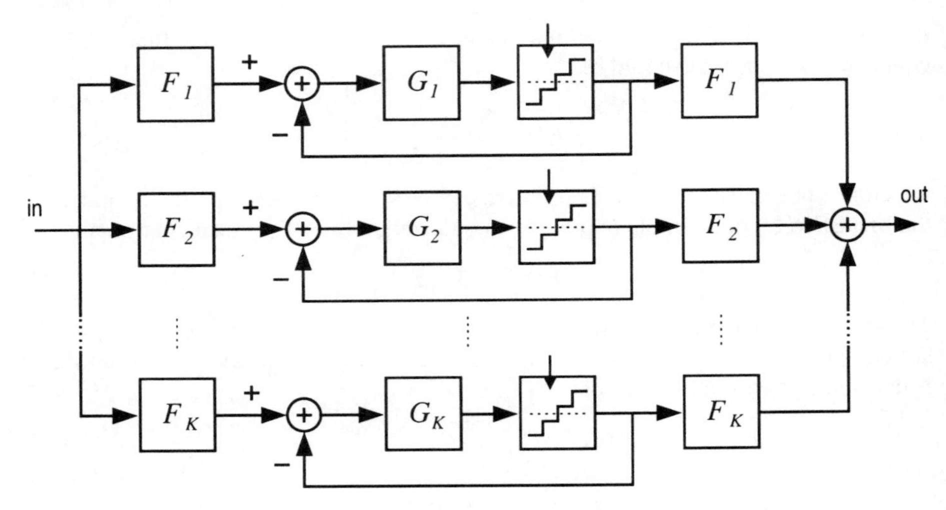

Figure 3.6: *Multi band parallel SDM.*

nels which are quantized using as many SDMs. Because each SDM has to suppress the quantization errors in a different part of the frequency spectrum, they all have different loop filters. In order to retrieve a quantized version of the input signal, the outputs of the SDMs are filtered and combined.

Another way for spectral separation of the individual channels is the modulation the input signal of each SDM with a so-called Hadamard sequence [21] (see Fig. 3.7). A Hadamard sequence is a repeated row of the Hadamard transform matrix which only contains the values $+1$ and -1. Modulating a signal with a Hadamard sequence only changes its sign in certain time intervals and can be implemented easily. As the Hadamard sequences are orthogonal, modulating the input signal with a different sequence moves a different part of its spectrum to the lower frequency range. As a result, each SDM can have the same loop filter but will quantize a different part of the spectrum of the input signal. The outputs of the SDMs are filtered, demodulated and combined to retrieve a quantized version of the input signal.

Parallelism of noise shapers or SDMs can also be achieved in the time domain by means of time-interleaved sampling [22]. Using the idea of block filter theory, the combined NTF of K identical, mutually coupled, SDMs sampled at f_s can be made

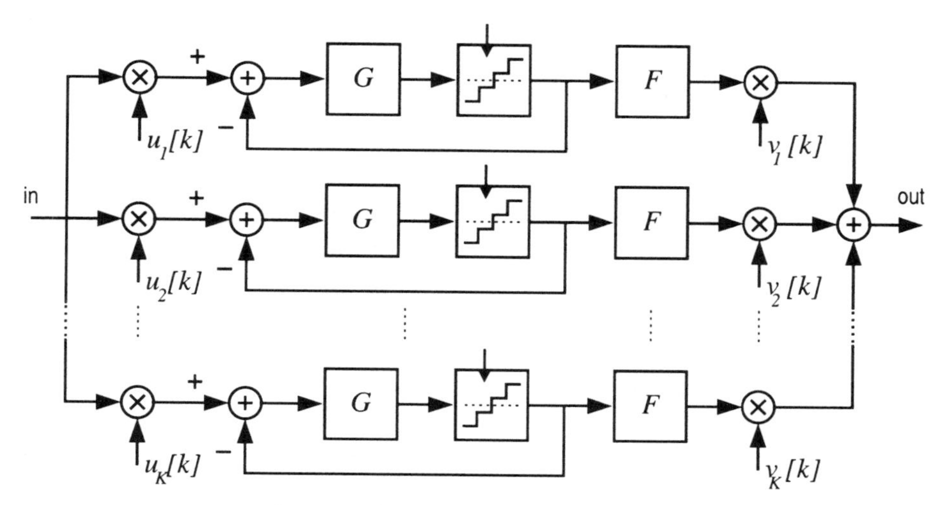

Figure 3.7: Hadamard parallel SDM.

equivalent to the NTF of a modulator running at Kf_s. As a result, the effective bandwidth in which the quantization errors are sufficiently suppressed is increased by a factor K, without changing the sample frequency of each individual SDM.

B. Reduced Bit Rate and Vector Quantization

In a conventional SDM, a quantizer is used that maps an input signal on a discrete output level, and a loop filter is used to predict the resulting quantization errors. In [23] the quantizer has been replaced by a general 1-to-m mapping which maps the current input to m outputs (see Fig. 3.8). The mapping effectively estimates the next m quan-

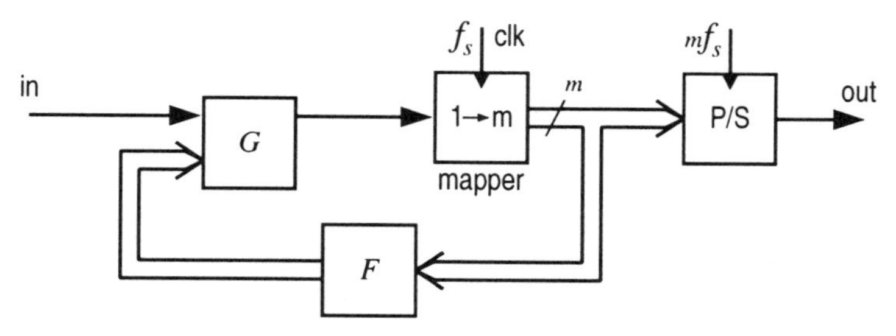

Figure 3.8: A reduced sample rate SDM.

tizer output levels of a conventional SDM, thereby reducing the sample rate with a factor of m. Because the mapping has m outputs, the loop filter is replaced by a multiple input, single output loop filter. The m parallel outputs are converted to a serial stream by the parallel to serial converter (P/S).

Instead of changing the number of outputs of the quantizer, the number of inputs can be increased [24]. The inputs are connected to several outputs of the loop filter, each representing a state variable of the filter. The resulting vector quantization does not decrease the sample rate, but aims to improve the stability and performance of the SDM.

C. Complex Bandpass SDMs

A very specific structure is the complex bandpass SDM [25]. In this modulator, shown in Fig. 3.9, a complex signal is sampled and quantized. Two real valued signals, corre-

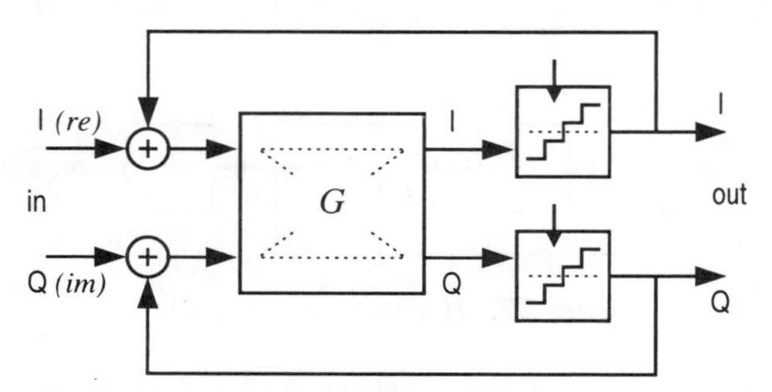

Figure 3.9: *A Complex (bandpass) SDM.*

sponding to the real and imaginary part, represent the complex signal. The two signals are usually referred to as I (in-phase) and Q (quadrature-phase) respectively. The real valued signals are quantized separately, but are mutually coupled through the complex loop filter. A key advantage of this structure is that the zeroes and poles implemented by the loop filter do not have to occur in complex conjugate pairs. As a result, the order of the loop filter for bandpass modulators can be reduced by a factor of two. Possible stability problems can be avoided at the cost of additional hardware.

3.4 Decimation and Filtering

The output of a noise shaper is a highly oversampled and roughly quantized signal. It not only represents the input signal but also contains the spectrally shaped quantization errors. Most of the quantization errors fall outside the signal bandwidth $[-f_b, f_b\rangle$ and can be removed by filtering (see Fig. 3.10). In the case of Analog-to-Digital conversion, the signal can be resampled at a lower frequency after the filtering operation. Because all the out-of-band signals are removed by the filter, the out-of-band quantization errors of the noise shaper will not alias into the signal band. The combined process of filtering and resampling for Analog-to-Digital conversion is called decimation.

Decimation is often performed using several stages and intermediate sampling frequencies. This is done in order to have an optimal trade off between complexity and power consumption. In the first stage, the sampling frequency is still very high, and simple filters are used to filter out a significant amount of quantization errors and reduce the sample frequency. The second stage consists of a more complex filter with a high attenuation outside the signal band and a very narrow transition region between

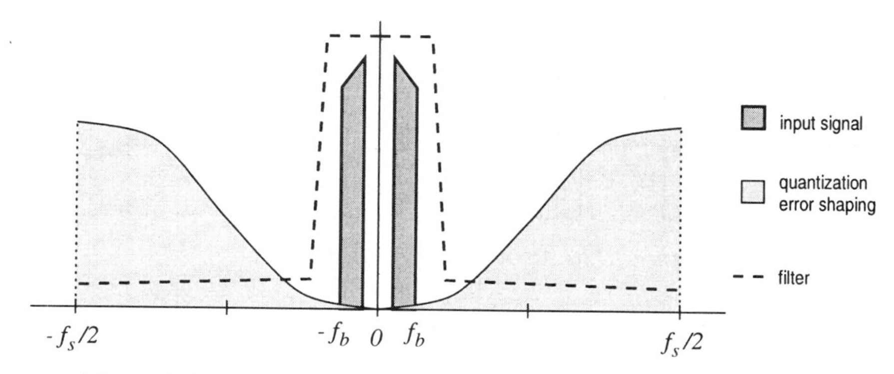

Figure 3.10: *Filter to remove the spectrally shaped quantization errors.*

the pass band and the stop band. This filter removes any spurious signals and quantization errors outside the signal band to allow resampling at the minimum (Nyquist) sampling frequency.

3.5 System Overview

Noise shaping and oversampling are used to increase the resolution of a coarse quantizer and improve the performance of ADCs and DACs. A typical analog-to-digital conversion using a noise shaper or SDM is shown in Fig. 3.11. The input filter re-

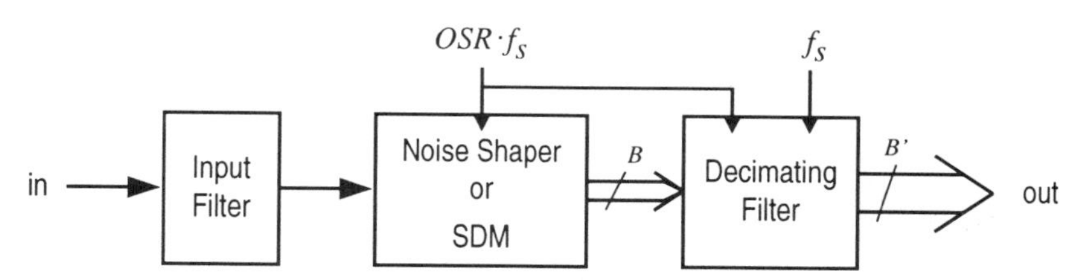

Figure 3.11: *An oversampling ADC system.*

moves spurious out-of-band signals from the analog input signal. The filtered signal is then fed into the oversampled noise shaper or SDM. The loop filter(s) of the noise shaper or SDM can be implemented using continuous time or discrete time elements. The digital output of the noise shaper or SDM consists of B bits representing the active output level of the quantizer. The decimating filter removes the out-of-band quantization errors using one or more stages. The digital output of the decimating filter consists of B' bits and is usually sampled at the Nyquist rate.

Similar to the oversampled ADC system, an oversampled noise shaper or SDM can be used for digital-to-analog conversion. A typical oversampled DAC system is shown in Fig. 3.12. The original sample rate f_s of the digital input signal is increased to $\mathrm{OSR} \cdot f_s$. In addition, the resampled input signal is filtered to remove any unwanted out-of-band signals. The noise shaper or SDM reduces the number of bits to B and

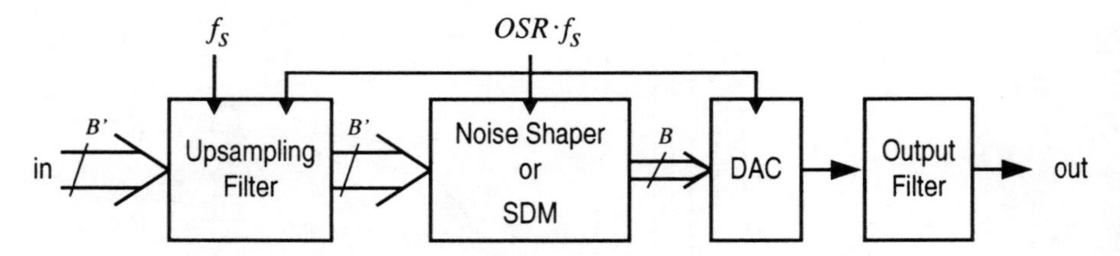

Figure 3.12: An oversampling DAC system.

suppresses the in-band quantization errors caused by the requantization of the input signal. As both the input and the output of the noise shaper or SDM are digital signals, the noise shaper or SDM can be implemented fully digitally. The output of the noise shaper or SDM is fed into a low resolution DAC. The resolution of the DAC is often equal to $B = 1$. The analog output of the DAC contains the input signal and shaped quantization noise (see Fig. 3.10). An analog filter with a high stop band suppression and a narrow transition region between the pass band and the stop band is needed to successfully remove the large out-of-band quantization errors.

3.6 Design Considerations

Apart from the architecture, several other considerations must be made when designing an oversampled error feedback coder. The performance of the coder not only depends on the order of the loop filter, the resolution of the quantizer and the oversampling ratio (see chapter 4), but also depends on the nonideal circuit behavior of the actual implementation. Another important design consideration is the stability of the nonlinear feedback loop.

3.6.1 Stability

The combination of a high order linear filter and a nonlinearity in a feedback loop can cause instability. The output and state of the system will no longer be controllable through the input. Due to the nonlinearity of the quantizer, the stability properties of the noise shaper or SDM depend on the applied input signal. Noise shapers or SDMs with a second or higher order loop filter are generally considered to be conditionally stable: the system will remain stable as long as the input signal and the state variables of the filter meet certain requirements. Unfortunately, stability analysis methods for linear systems do not apply and the conditions for which a noise shaper or SDM is stable have not yet been determined analytically. The stability of the feedback loop of the noise shaper and SDM will be investigated extensively in chapter 5.

3.6.2 Loop Filter Topologies

In the literature concerning noise shapers and SDMs, a considerable amount of attention is given to specific loop filter topologies [2]. This is probably caused by the fact that lowpass SDMs historically have built with an interpolative loop filter structure, consisting of a chain of integrators. One of the first SDMs with a loop filter order higher than one was constructed using a "double loop" [26]. In order to increase the performance or affect the stability properties of the SDM, multiple feedforward paths and local resonator feedbacks were applied, resulting in the loop filter topology shown in Fig. 3.13.

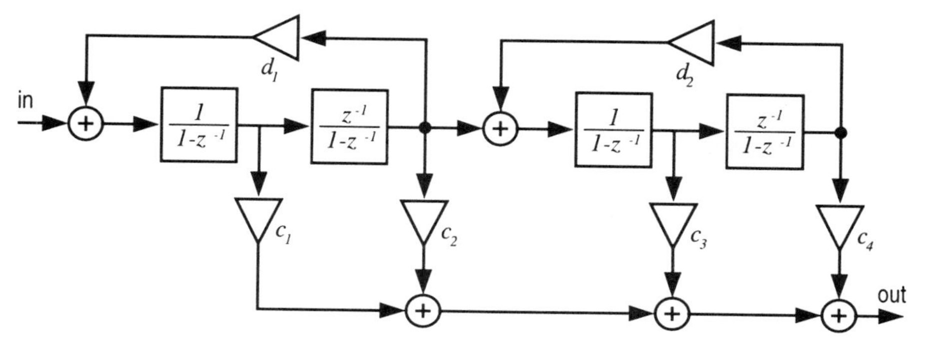

Figure 3.13: Interpolative loop filter topology.

Although for practical reasons these topologies can be used for the implementation of the loop filter, they are a special case and can be transformed into the direct form shown in Fig. 3.14 [9]. The transfer function of this discrete time filter equals

$$G(z) = \frac{G_c(z)}{G_d(z)} = \frac{c_0 + c_1 z^{-1} + c_2 z^{-2} + \cdots + c_N z^{-N}}{1 - d_1 z^{-1} - d_2 z^{-2} - \cdots - d_N z^{-N}} \tag{3.26}$$

in which N is the order of the loop filter, c_n $(n = 0, 1, \ldots, N)$ are the feedforward coefficients determining the zeroes of $G(z)$ and d_n $(n = 1, 2, \ldots, N)$ are the feedback coefficients determining the poles of $G(z)$. When placed inside the loop of a noise shaper or SDM, the loop filter transfer function should have a delay of at least one sample period in order to result in a causal (and hence realizable) system. Consequently, the coefficient c_0 in Fig. 3.14 should be $c_0 = 0$.

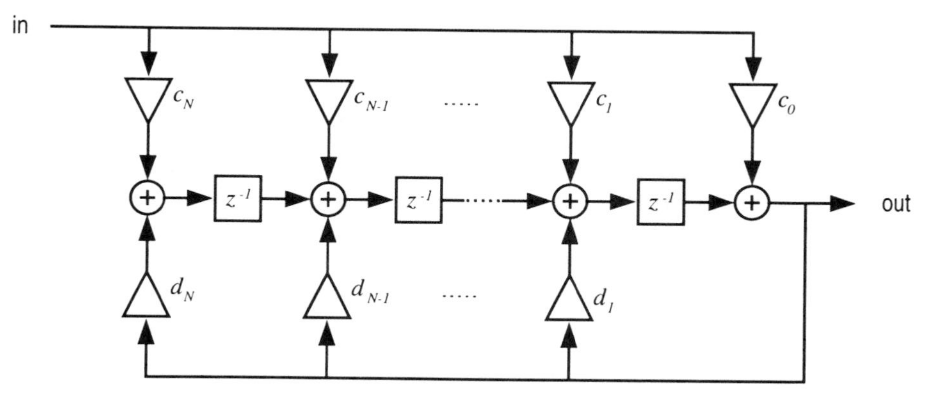

Figure 3.14: Direct form filter topology.

3.6.3 Implicit Input Filtering

A major disadvantage of the direct feedback structure of the SDM in Fig. 3.4 is that both the signal transfer function STF and the noise transfer function NTF solely depend on the loop filter $G(z)$. Designing the NTF directly determines the STF (see eq. (3.9)). In order to allow additional freedom in the design of the STF, the loop filter of Fig. 3.14 could be extended with separate feed forward coefficients $c_{n,i}$ and $c_{n,o}$ for the SDM input i and the SDM output o, as is shown in Fig. 3.15. The resulting SDM structure is shown in Fig. 3.16. The output $O(z)$ of the SDM can be written as

$$O(z) = \frac{G_{c,i}(z)}{G_{c,o}(z)} \cdot \frac{G(z)}{1+G(z)} \cdot I(z) + \frac{1}{1+G(z)} \cdot E_q(z) \tag{3.27}$$

with $G(z) = G_{c,o}(z)/G_d(z)$ as defined by (3.26). Compared to the expression for the SDM output in eq. (3.9), the input is additionally filtered with $G_{c,i}(z)/G_{c,o}(z)$. This provides additional degrees of freedom in the design of the zeroes of the STF. This can be easily seen by substituting $G(z) = G_{c,o}(z)/G_d(z)$. The original STF can be written as:

$$\text{STF}(z) = \frac{G(z)}{1+G(z)} = \frac{G_{c,o}(z)}{G_{c,o}(z)+G_d(z)} \tag{3.28}$$

By adding the additional feedforward coefficients the zeroes of the STF are replaced:

$$\text{STF}(z) = \frac{G_{c,i}(z)}{G_{c,o}(z)} \cdot \frac{G(z)}{1+G(z)} = \frac{G_{c,i}(z)}{G_{c,o}(z)+G_d(z)} \tag{3.29}$$

The additional freedom in the placement of the zeroes of the STF can be used to suppress unwanted signals in the input of the SDM at the expense of some extra hardware. An example of the application of implicit input filtering is given in section 7.3.5.

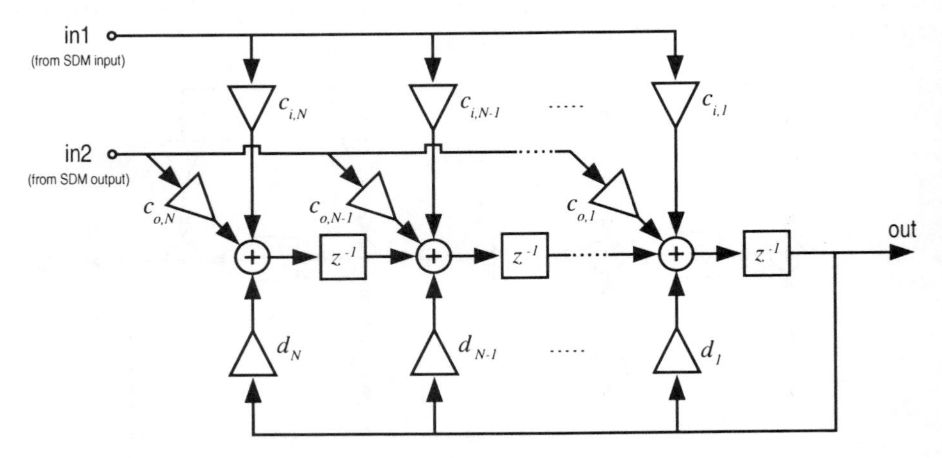

Figure 3.15: *Direct form filter topology with two inputs.*

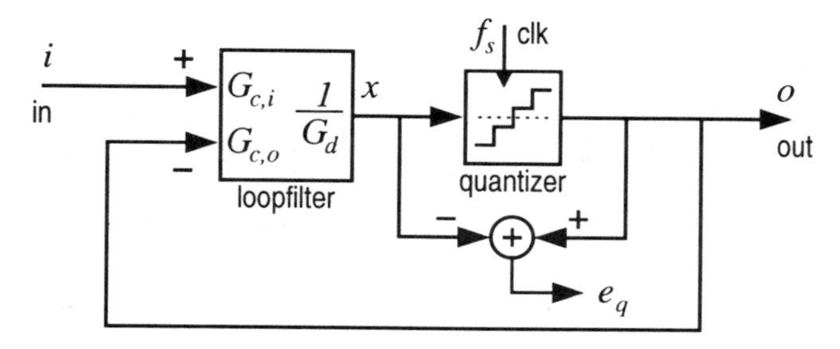

Figure 3.16: SDM with input filtering.

3.6.4 Continuous-time vs. Discrete-time Loop Filters

Until now, the signals and the loop filter of a noise shaper or SDM have been considered to be discrete time. Although this simplifies the understanding of the noise shaping concepts, it is not a requirement. The loop filter can also be implemented using continuous time elements. As a result, the input and output signals of the loop filter will be continuous time signals. As the output of the SDM will remain a discrete time signal, a discrete to continuous time conversion needs to be performed in the feedback loop using a DAC. Figure 3.17 shows the resulting SDM with a continuous time loop filter. Continuous time signals are indicated using the continuous time index t, e.g. $i(t)$,

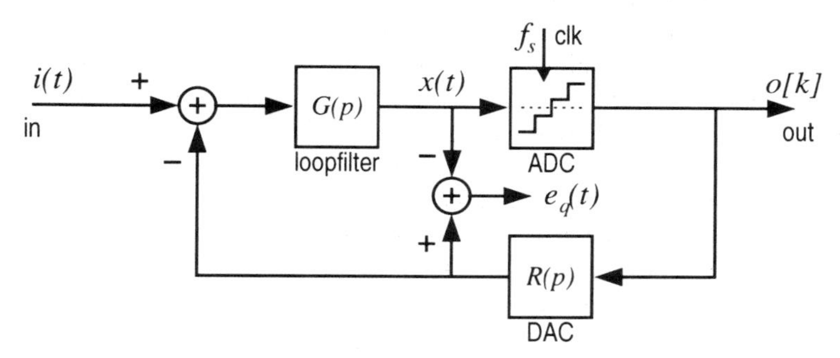

Figure 3.17: SDM with continuous time loop filter.

and discrete time signals are indicated using the discrete time index k, e.g. $o[k]$. Similarly, the Laplace transform $G(p)$ is used to represent the continuous time loop filter transfer function. The continuous time transfer function of the DAC in the feedback loop is indicated by $R(p)$.

Although the loop filter is continuous time, the overall behavior of the SDM can be described by discrete time transfer functions [27]. The loop filter and the DAC can be replaced by an equivalent discrete time filter $G(z)$, which depends on the continuous time loop filter transfer function $G(p)$, the DAC transfer function $R(p)$ and the sample frequency f_s. The equivalent discrete time counterpart of the continuous time SDM is shown in Fig. 3.18. As the continuous time input signal $i(t)$ is not sampled before it enters the loop filter $G(p)$ in Fig. 3.17, a part of the filtering operation will not be

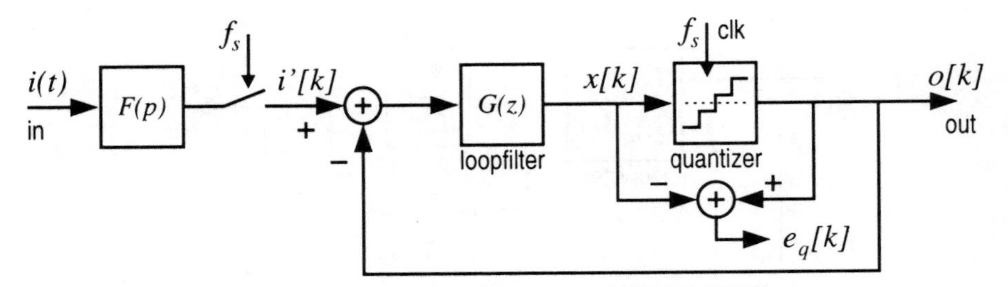

Figure 3.18: *Equivalent discrete time SDM of Fig. 3.17.*

replaced by the discrete time loop filter and some residual analog filtering has to be performed before the input signal is sampled.

A key advantage of using a continuous time loop filter instead of a discrete time loop filter is that the sampling operation takes place inside the loop. As a result, additional errors due to sampling or spurious out-of-band signals at the input of the quantizer are also suppressed by the feedback loop.

3.6.5 One-bit vs. Multi-bit Quantizers

Another consideration when designing a noise shaper or similar feedback coder is the resolution of the quantizer. The overall performance not only depends on the number of quantization levels, but also depends on the accuracy of the quantizer implementation. The accuracy of the individual levels of the quantizer determines the final performance that can be achieved by the noise shaper. For example, when a four-bit quantizer is used in an oversampled noise shaper to achieve an ENOB of 16 bit, the individual levels of the quantizer should have at least 16 bit accuracy. This requirement complicates the design of the quantizer considerably.

An exception to this requirement is the one bit quantizer. Any mismatch in either of the two quantization levels merely results in a DC offset and/or a gain mismatch (see Fig. 3.19). As these two errors are linear deviations of the ideal operation of the quantizer, they do not degrade the performance of the noise shaper. The effects of non-ideal implementation of the quantizer will be discussed more in detail in section 4.4.

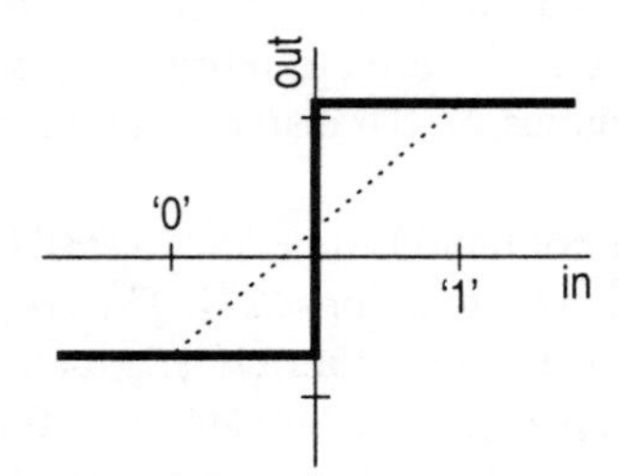

Figure 3.19: *Level mismatch in a one bit quantizer.*

3.7 Conclusions

In this chapter, the basic concepts of oversampling and noise shaping have been presented. Several architectures for noise shaping were discussed and a few design considerations were highlighted. The rest of this book will focus mainly on single loop, higher order one-bit SDMs. This class of error feedback coders can achieve a high SNR at moderate oversampling ratios (see chapter 4) and contains circuitry that is relatively easy to design[1].

The major obstacle in the design of single loop, high order one-bit modulators is the issue of input signal dependent stability. The stability of these SDMs will be investigated and discussed in chapter 5. Chapter 6 will focus on the impact of the stability requirements on the design of SDMs and bandpass modulators in particular.

[1] A higher order one-bit SDM consists of a single continuous time or discrete time loop filter, a comparator and a one-bit DAC.

CHAPTER 4

PERFORMANCE

The performance of oversampled sigma delta modulators can be predicted by modeling of the quantization errors as additive white noise. The theoretically achievable performance depends on several parameters such as the oversampling ratio, the order of the loop filter and the resolution of the quantizer. Deterministic effects such as pattern noise and dead zones, as well as a non-ideal implementation affect the actual performance of an SDM.

4.1 Linear Prediction

In section 3.3.1 the output of a single loop sigma delta modulator was calculated. The output can be written in terms of the input signal $I(z)$, the loop filter $G(z)$ and the quantization error signal $E_q(z)$:

$$O(z) = \frac{G(z)}{1+G(z)} \cdot I(z) + \frac{1}{1+G(z)} \cdot E_q(z) \tag{4.1}$$

In the case that the quantization errors could be regarded as additive white noise, as suggested in section 2.3.1, the output of the SDM would consist of a filtered input signal and a shaped noise spectrum. The signal to noise ratio (SNR) in the band of interest could be calculated easily [28]. Unfortunately, the quantizers used in SDMs hardly ever match the requirements to validate the white noise approximation:

- The number of quantization levels is usually very small.
- The quantizer can be in overload.
- The input signal of the quantizer is generally not a random (white) signal.

Although adding a dither signal could satisfy the latter requirement, the low resolution of the quantizers renders the model invalid. A one-bit quantizer in particular cannot be modeled by the addition of independent noise. The output power of a one-bit quantizer is constant. Adding quantization noise with constant power to an input signal with variable power does not model the quantizer accurately. In order to accommodate the variable input, the quantizer can be modeled by an additive white noise source n_q and a time invariant gain c_g [29]. The resulting model is shown in Fig. 4.1. In the case of a one-bit quantizer, the gain c_g can have any value larger than zero. For multi-bit quantizers the value of c_g will be near unity. The quantization noise n_q has a total power of $q^2/12$ and is considered to be spectrally white and independent of the

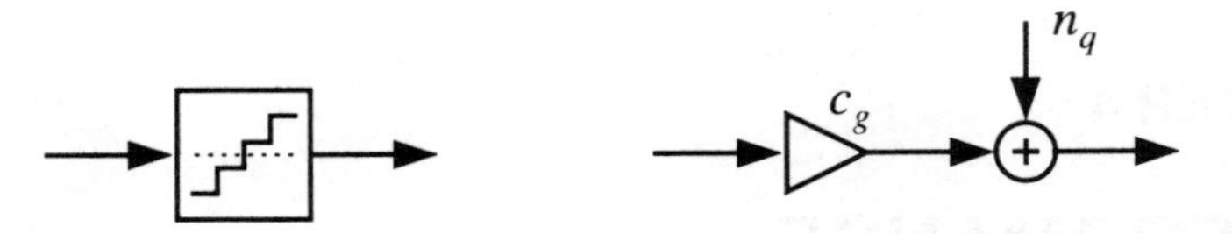

Figure 4.1: *Modified quantizer noise model.*

quantizer input signal. In z-domain, the quantization noise power density is given by:

$$N_q(e^{j\theta}) = \frac{q^2}{24\pi} \tag{4.2}$$

Applying this model to the quantizer in an SDM results in the linear noise prediction model shown in Fig. 4.2. The output of this linear model equals

$$\tilde{O}(z) = \frac{c_g G(z)}{1+c_g G(z)} \cdot I(z) + \frac{1}{1+c_g G(z)} \cdot N_q(z) \tag{4.3}$$

Compared to eq. (3.9) the STF and the NTF now include the noise parameter c_g and will be referred to as the modified STF and modified NTF respectively. For $c_g = 1$ the original STF and NTF are obtained. If all the signals are assumed to be real-valued, the noise contribution for negative frequencies will be identical to the contribution for positive frequencies. The in-band noise contribution N_o in the output of the quantizer can then be calculated by

$$N_o = 2 \cdot \int_{\theta_{\min}}^{\theta_{\max}} \left| \frac{1}{1+c_g G(e^{j\theta})} \right|^2 \cdot \frac{q^2}{24\pi} \, d\theta \tag{4.4}$$

where $[\theta_{\min}, \theta_{\max}]$ is the angular frequency range (bandwidth) of the input signal. In the case of a multi-bit quantizer, the value for the gain will be considered unity: $c_g = 1$. For a one-bit quantizer, the value for c_g can be determined by calculating the total output power. As the output of a one bit quantizer is either $-q/2$ or $q/2$, the output power is constant and equal to $q^2/4$. The total output power of the linear model according to (4.3) should therefore satisfy

$$\int_{-\pi}^{\pi} \left| \frac{c_g G(e^{j\theta})}{1+c_g G(e^{j\theta})} \right|^2 |I(e^{j\theta})|^2 \, d\theta + \int_{-\pi}^{\pi} \left| \frac{1}{1+c_g G(e^{j\theta})} \right|^2 \frac{q^2}{24\pi} \, d\theta = \frac{q^2}{4} \tag{4.5}$$

If the gain of the loop filter $G(z)$ is very large within the bandwidth of the input signal, the transfer from the input signal to the output of the SDM will be unity. The output

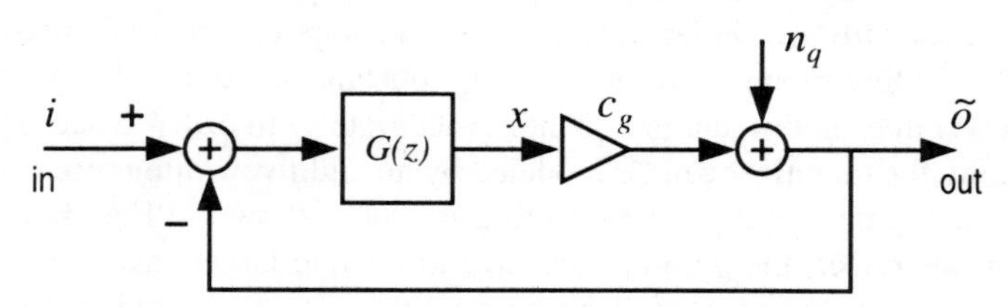

Figure 4.2: *Linear noise prediction model for an SDM.*

power will be equal to the total noise power plus the input signal power P_i, and (4.5) can be simplified to

$$P_i + \int_{-\pi}^{\pi} \left| \frac{1}{1 + c_g G(e^{j\theta})} \right|^2 \frac{q^2}{24\pi} \, d\theta = \frac{q^2}{4} \tag{4.6}$$

from which c_g can be solved. For $P_i = 0$ the resulting value for c_g can be can be used to calculate the idle-channel in-band noise power with (4.4). In order to determine the dynamic range (DR) of the modulator, the maximum signal power has to be calculated. In section 2.5 it was shown that the maximum sine wave amplitude of a B bits quantizer can be approximated by:

$$A_{\max} = (2^B - 2 + \frac{4}{\pi})q/2 \tag{4.7}$$

resulting in a maximum signal power at the output of the quantizer:

$$P_{i,\max} = \frac{A_{\max}^2}{2} = \left(2^B - 2 + \frac{4}{\pi}\right)^2 q^2/8 \tag{4.8}$$

The DR of the modulator is found by dividing the maximum signal power $P_{i,\max}$ of (4.8) by the idle-channel noise power $N_{o,idle}$ that was obtained using (4.4) and (4.6). The value for the dynamic range can be used as an estimate for the maximum signal-to-noise ratio (SNR). In [30] STIKVOORT showed that this estimate is too high. The value of c_g decreases with an increasing input power, thereby changing the spectral shaping of the quantization noise. As a result, the in-band noise power increases with an increase of the input amplitude.

Although this model proved to be considerably accurate for higher order modulators [29], it failed to accurately predict the idle channel quantization noise of a first order modulator. A more elaborate model for calculating the quantization noise was developed by ARDALAN and PAULOS. In [31] they suggested to use a separate gain for the signal and the quantization noise. By doing so, the correlation between the input of the quantizer x and the quantization noise n_q is incorporated in the model. In the model of Fig. 4.2 the input of the quantizer and the quantization errors are assumed to be uncorrelated. As the model of ARDALAN and PAULOS uses two parameters, these parameters cannot be solved uniquely through a power requirement such as (4.5). Instead, the two parameters are solved by minimizing the rms error in modeling the nonlinear quantizer. Although fitting a more elaborate model to the non-linear quantizer can result in more accurate predictions, it does not give significantly more insight in the behavior of the SDM.

4.1.1 Lowpass Modulator Example

As an example, the DR of a set of lowpass modulators with a one bit quantizer is calculated. The prototype loop filter of the modulator equals

$$G(z) = \frac{(1 - az^{-1})^N}{(1 - bz^{-1})^N} - 1 \tag{4.9}$$

in which b is the radius of the poles (usually $b = 1$), a determines the position of the zeroes and N is the order of the loop filter. The modified NTF of the modulator equals

$$\mathrm{NTF}(z) = \frac{(1-bz^{-1})^N}{(1-c_g)(1-bz^{-1})^N + c_g(1-az^{-1})^N} \tag{4.10}$$

In the ideal case that $c_g = 1$, the (modified) NTF is equal to

$$\mathrm{NTF}(z) = \frac{(1-bz^{-1})^N}{(1-az^{-1})^N} \tag{4.11}$$

The effects of the noise shaping of these modulators can be maximized by placing the zeroes of the NTF on $z = 1$ by choosing $b = 1$, and placing the poles of the NTF as far away as possible. For a first and second order modulator, the poles can be placed in the origin by choosing $a = 0$. In the case of high order modulators ($N > 2$), the poles have to be placed closer to the zeroes in order for the modulator to be stable for small input signals (see chapter 5). Table 4.1 lists the values for the filter parameters and the resulting idle channel value for c_g. The dynamic range of these modulators is shown in Fig. 4.3 and Fig. 4.4 as a function of the oversampling ratio OSR. For very low oversampling ratios (OSR < 4) a first order modulator achieves the highest dynamic range. When the OSR lies between OSR ≈ 4 and OSR ≈ 12 a second order modulator optimizes the DR. The requirements on the NTF pole placement for higher order modulators degrades their performance for such low oversampling ratios. Only when OSR > 30 will an increase of the order of the loop filter also result in an increase in the DR of the modulator.

4.1.2 Optimal NTF zero placement

The loop filters defined by (4.9) place all the NTF zeroes (suppressing the quantization noise) at the same frequency. For these lowpass modulators the zeroes are located at DC ($z = 1$). In most cases, this is not the solution which will minimize the in-band quantization noise of (4.4). For low to moderate oversampling ratios in particular, the NTF zeroes should be spread across the signal band. An estimate for the optimal placement of the NTF zeroes of a lowpass modulator can be calculated by minimizing

$$\int_0^{\theta_b} \left|\Pi_{l=0}^{N}(z - e^{\theta_l})\right|^2 \, d\theta \qquad \text{with} \qquad z = e^{j\theta} \tag{4.12}$$

in which θ_b is the normalized bandwidth of the signal and θ_l ($l = 0 \ldots N$) are the normalized angular frequencies of the zeroes. In order to have a loop filter with real valued coefficients, complex zeroes should occur in conjugate pairs. For example, the NTF zeroes of a second order modulator should be placed at $z = e^{\pm j\theta_0}$. The angular frequency θ_0 can be determined by calculating

$$\min_{\theta_0} \int_0^{\theta_b} \left|(z - e^{j\theta_0})(z - e^{-j\theta_0})\right|^2 \, d\theta \qquad \text{with} \qquad z = e^{j\theta} \tag{4.13}$$

Table 4.1: *Filter parameters and the idle channel value for c_g used to determine the DR of a one bit SDM with a loop filter according to (4.9).*

N	a	b	c_g	N	a	b	c_g
1	0	1	1.3333	4	0.619	1	0.8458
2	0	1	0.6667	5	0.717	1	0.9197
3	0.416	1	0.7166	6	0.771	1	0.9400

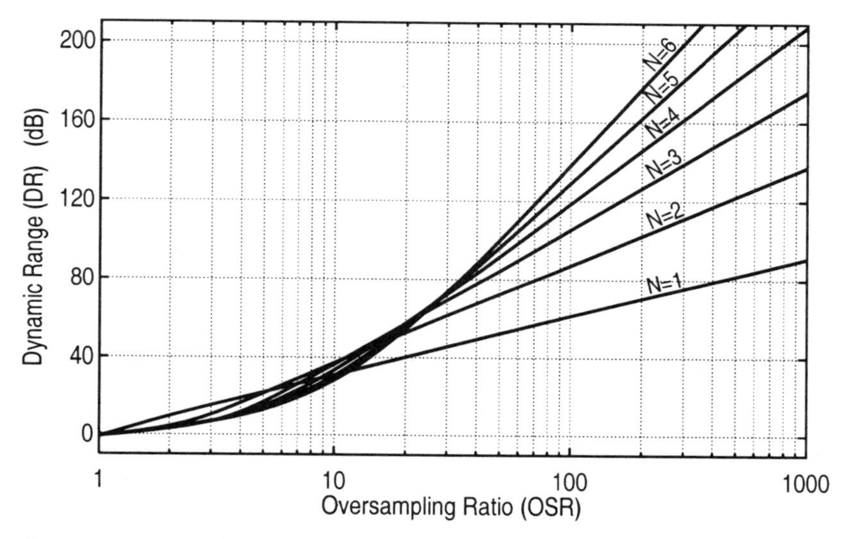

Figure 4.3: *Dynamic range of a one bit SDM with loop filter (4.9). The corresponding filter parameters are listed in table 4.1.*

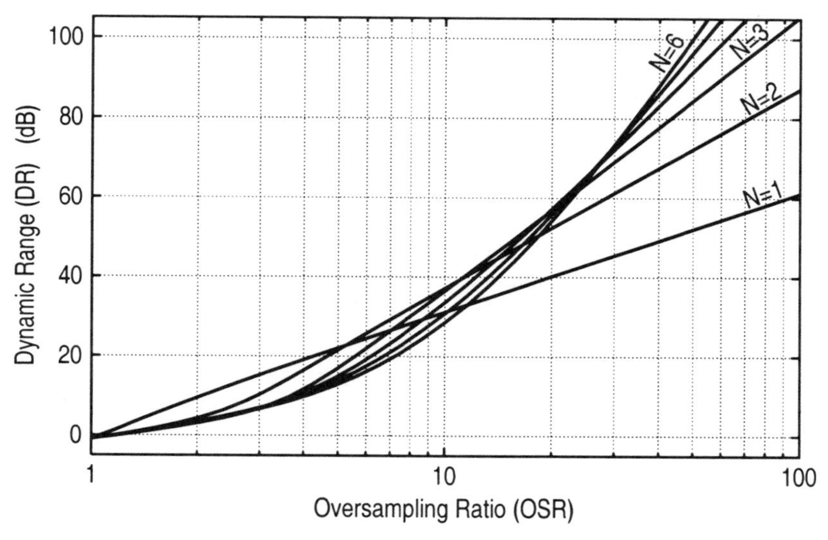

Figure 4.4: *Detail of Fig. 4.3.*

Evaluating the integral results in

$$\min_{\theta_0} 2\left\{\theta_b(2+\cos 2\theta_0)+\frac{1}{2}\sin 2\theta_b - 4\sin\theta_b\cos\theta_0\right\} \tag{4.14}$$

from which the optimal solution for the zero placement can be solved:

$$\theta_0 = \arccos\left(\frac{\sin\theta_b}{\theta_b}\right) \overset{\theta_b \ll 1}{\approx} \sqrt{\tfrac{1}{3}}\cdot\theta_b \tag{4.15}$$

The relocation of the NTF zeroes of the second order modulator increases the DR with approximately 3.5dB compared to placing both the zeroes at $z = 1$. Similar calculations can be done for higher order modulators. For modulators with a loop filter order of $N > 4$ in particular, a relocation of the NTF zeroes across the signal band results in a significant increase in the DR of the modulator ([2], p.154). Recently, a combined theoretical and empirical study on influence of the loop filter parameters of lowpass modulators on the performance was presented in [32].

4.2 Idle Patterns, Dead Zones and Tones

The linear prediction method in the previous section provides an estimate for the in-band quantization noise and the resulting DR of a modulator. The linear modeling fails to predict certain aspects of the behavior of the SDM. As the modulator is sampled, the output signal and the quantization errors often contain repeating patterns causing spurious tones in the frequency spectrum and other nonlinear behavior.

4.2.1 Idle Patterns and Dead Zones

The repetitive patterns that are present in the output of the modulator under zero input conditions are called idle patterns. They are the result of limit cycles (see Chapter 5). Depending on the loop filter and the initial state of the modulator, the actual idle pattern may vary. Typical idle patterns of lowpass one bit modulators include repeating series of '10', '1100' or '11010010' patterns in which a '1' corresponds to a quantizer output value of $q/2$ and a '0' corresponds to $-q/2$. One of the effects of these patterns is a limited input sensitivity. For input signals with a very small amplitude, the idle patterns in the output of the modulator are not disturbed. The small input signal will not be coded by the SDM, resulting in a so-called "dead-zone".

The dead-zone of a first order lowpass modulator with a loop filter $G(z) = z^{-1}/(1-z^{-1})$ can be calculated easily. Without an input signal applied to the modulator ($i[k] = 0$), the quantizer input signal and modulator output signal contain an idle pattern with frequency $f_s/2$ (shown in Tab. 4.2). The parameter x_0 is the initial state of the filter and has a value $0 < x_0 < q/2$. When a small input signal is applied to the modulator, the sum (or integral) of the input samples adds to the quantizer input. When the absolute value of this integral exceeds either x_0 or $q/2 - x_0$, the output of the quantizer changes state and the idle pattern is disturbed. The sensitivity of the idle pattern depends on the

Table 4.2: Idle pattern of a first order lowpass modulator.

k	0	1	2	3	4	5
$o[k]$	$\frac{q}{2}$	$-\frac{q}{2}$	$\frac{q}{2}$	$-\frac{q}{2}$	$\frac{q}{2}$	$-\frac{q}{2}$
$x[k]$	x_0	$x_0-\frac{q}{2}$	x_0	$x_0-\frac{q}{2}$	x_0	$x_0-\frac{q}{2}$

initial state x_0 and is worst for $x_0 = 0$ or $x_0 = q/2$. The threshold value for the sum of the input samples then equals $q/2$. In the case that the input signal is a sine wave

$$i[k] \;=\; A_i \cdot \sin(2\pi f_i k T_s) \tag{4.16}$$

with $T_s = 1/f_s$, the sum of the input samples can be approximated by

$$\sum_{k=0}^{K} i[k] \;\approx\; \frac{1}{T_s}\int_{t=0}^{KT_s} A_i \cdot \sin(2\pi f_i t)\, dt \;=\; \frac{A_i f_s}{2\pi f_i} \cdot \{\cos(2\pi f_i K T_s) - 1\} \tag{4.17}$$

when $K \gg 1$ and $f_s \gg f_i$. The sum of the input samples has an absolute maximum of

$$\left|\sum_{k=0}^{K} i[k]\right|_{\max} \;\approx\; \frac{A_i f_s}{\pi f_i} \tag{4.18}$$

In the case that this absolute maximum does not exceed the value of $\frac{q}{2}$, the idle pattern will not be disturbed. The first order lowpass modulator has a dead-zone for sine waves with frequency f_i and an amplitude A_i that satisfies

$$A_i < \frac{q\pi f_i}{2 f_s} \tag{4.19}$$

The dead-zone of a first-order SDM is most noticeable for input signals with a relatively high (in-band) frequency. Modulators with a higher order loop filter also exhibit a dead-zone which can be calculated in a similar manner. Because of the higher gain in the loop filter, the dead-zone of these modulators will be considerably smaller than the dead-zone of the first order SDM. Very small signals (including system noise) will disturb the idle patterns of such SDMs, rendering the dead-zone undetectable in practical implementations.

4.2.2 Tones

The same mechanism that causes the idle patterns is also responsible for the pattern noise or tones. Depending on the input signal, the output of the modulator contains repetitive patterns, causing pattern noise or tones in the frequency spectrum. These tones are a considerable problem in audio or video coding applications. As the human ear and eye are very susceptible to these highly repetitive signals, the pattern noise and tones can be audible or visible, even when the system noise well exceeds the distortion power. The tones are also responsible for deviations in actual SNR performance: in-band tones cause bumps and slope changes in the SNR vs. input power characteristics.

The pattern noise and tones are most evident in first order and second order SDMs, but are also present in higher order modulators. As an example, Fig. 4.5 shows the in-band spectrum of a third order lowpass modulator with a DC input. The power of the DC input signal is -60dB compared to the quantizer output power $q^2/4$. The constant input level causes the SDM to produce repetitive output patterns, resulting in spurious tones at frequencies which are multiple of $f_s/1000$.

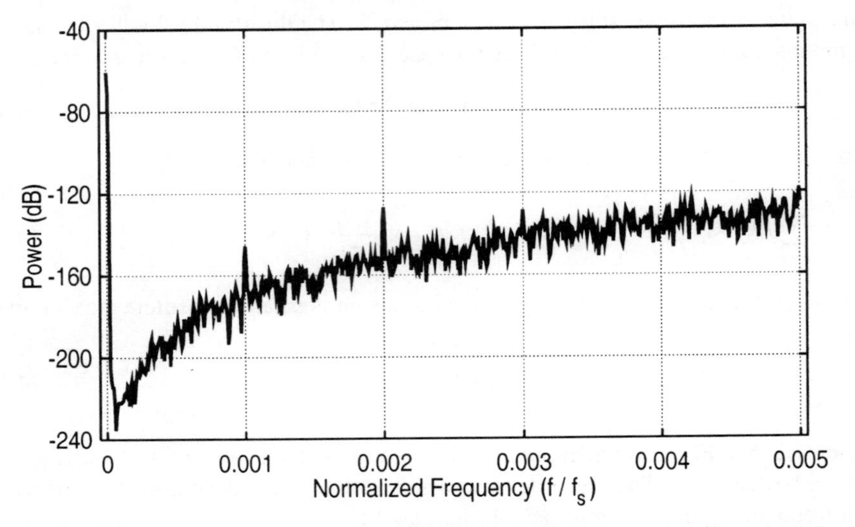

Figure 4.5: *In-band power spectral density of a third order lowpass modulator with DC input (simulation, 64K bins).*

The nature of the quantization errors of a first and second order lowpass modulator has been investigated extensively for various types of input signals [33–38]. The analysis methods used in these articles provide very accurate predictions for quantization noise spectra of the modulators. However, the complexity of the analytical methods renders them unsuitable for analyzing the quantization noise of high ($N \geq 3$) order modulators. Therefore, an approximate method will be presented hereafter. The method gives a qualitative insight in the mechanism that causes tones and the effects on the performance of the SDM. For reasons of simplicity, the analysis will be limited to SDMs with a one-bit quantizer. First, the tones in a first order modulator will be analyzed. The findings will then be applied to second and higher order modulators.

First Order Lowpass Modulator

The behavior of a first order lowpass SDM as defined by (3.14) will be analyzed under DC input bias conditions. Without loss of generality, the quantizer step size can be set to $q = 2$, resulting in quantizer output values of $\pm$ 1. The first order lowpass modulator can be described in discrete time domain by

$$x[k] = x[k-1] + i[k-1] - o[k-1] \tag{4.20}$$

$$o[k] = \mathrm{sgn}(x[k]) \tag{4.21}$$

in which $x[k]$ is the quantizer input, $i[k]$ is the modulator input and $o[k]$ is the modulator output (see Fig. 3.4). The initial state of the modulator is defined by $x[0] = x_0$ and $o[0] = \text{sgn}(x_0)$. Furthermore, let the input $i[k]$ for $k \geq 0$ be a rational fraction m/M, with m, M positive integers and $\gcd(m,M) = 1$. By substitution it can be shown that for $k \geq 1$ the input of the quantizer $x[k]$ can be written as

$$x[k] = \tfrac{m}{M} - \tfrac{1}{2} + \tfrac{1}{2}(-1)^{k+\lfloor \frac{(k-1)m}{M} + x_0 \rfloor} + \tfrac{(k-1)m}{M} + x_0 - \lfloor \tfrac{(k-1)m}{M} + x_0 \rfloor \qquad (4.22)$$

in which $\lfloor y \rfloor$ represents the largest integer smaller than y. The quantizer input $x[k]$ contains a constant, an alternating part corresponding to the idle pattern and a fractional part $y - \lfloor y \rfloor$ depending on the input signal and the initial state. The series described by (4.22) is periodic for k, with period M if m and M are odd, and $2M$ if either m or M is even. The output of the modulator, equal to the sign of $x[k]$, will have the same period. In the case that both m and M are odd, the output will change sign $(M - m)$ times in the period of M samples. As a result, the output will have an average (DC) value of $\frac{m}{M}$ as required, but will also contain spectral components with fundamental frequencies

$$f_L = \frac{m}{2M} f_s \qquad \text{and} \qquad f_H = \frac{M-m}{2M} f_s \qquad (4.23)$$

In the case that either m or M is even, the same results are obtained. The first frequency f_L corresponds to in-band tones as shown in Fig. 4.5. The latter frequency f_H shows that a DC input also results in tones near half of the sampling frequency. Although this fundamental frequency does not fall within the signal bandwidth, its harmonics can have an impact on the SNR. Additionally, these tones may result in in-band distortion if any crosstalk of clock signals is present within a practical implementation. The crosstalk can result in intermodulation, and the tones will shift to the baseband frequencies. Note that the power of the distortion component f_H will be considerably larger than the power of the distortion at f_L if m is small compared to M. In that case $(M - m)$ pulses (or sign changes) of the output add to the power of f_H and only m pulses add to the power of distortion components f_L.

The frequency of the distortion components which are present in the output of the first order lowpass modulator with a DC input depends on the DC input value. Although this relationship was derived for input values which are a rational fraction m/M, it will assumed to be valid for any DC input value. In the case of an irrational input value, the loop filter output $x[k]$ will be quasi-periodic and will result in an output containing repetitive patterns. With a DC input value of A_i, the frequencies of the distortion components can be expressed as:

$$f_{H,n} = n \cdot \left(1 - \tfrac{2A_i}{q}\right) \frac{1}{2} f_s \qquad \text{with } n = 1,2,3\ldots \qquad (4.24)$$

$$f_{L,n} = n \cdot \left(\tfrac{2A_i}{q}\right) \frac{1}{2} f_s \qquad \text{with } n = 1,2,3\ldots \qquad (4.25)$$

In the case that A_i is small compared to q, the components $f_{L,n}$ can be neglected. Because frequencies higher than $f_s/2$ will fold back (aliasing), the expression for the distortion frequencies $f_{H,n}$ can be changed to:

$$f_{Ho,n} = \left(1 - n \cdot \tfrac{2A_i}{q}\right) \frac{1}{2} f_s \qquad n \text{ odd} \tag{4.26}$$

$$f_{He,n} = n \cdot \left(\tfrac{2A_i}{q}\right) \frac{1}{2} f_s \qquad n \text{ even} \tag{4.27}$$

For $A_i = 0$, the distortion components at $f_{Ho,n}$ represent the idle frequency $f_s/2$. In the case that the input signal is a slowly varying signal, the idle pattern will appear to be frequency modulated with the input signal. Similar distortion signals will appear in the base band of the output signal due to $f_{He,n}$. The maximum deviation or "bandwidth" of these frequency modulated distortion signals can be calculated. In the case that the input signal is a sine wave $A_i \sin(2\pi f_i t)$, the bandwidth of the distortion signals $f_{H,n}$ is equal to $n \cdot f_D$, with:

$$f_D = \frac{A_i}{q} \cdot f_s \tag{4.28}$$

Figure 4.6 shows a typical spectrum of a first order lowpass modulator with a low frequency sine wave input. Apart from the input signal near DC, the spectrum contains frequency modulated distortion components with bandwidths that are a multiple of f_D.

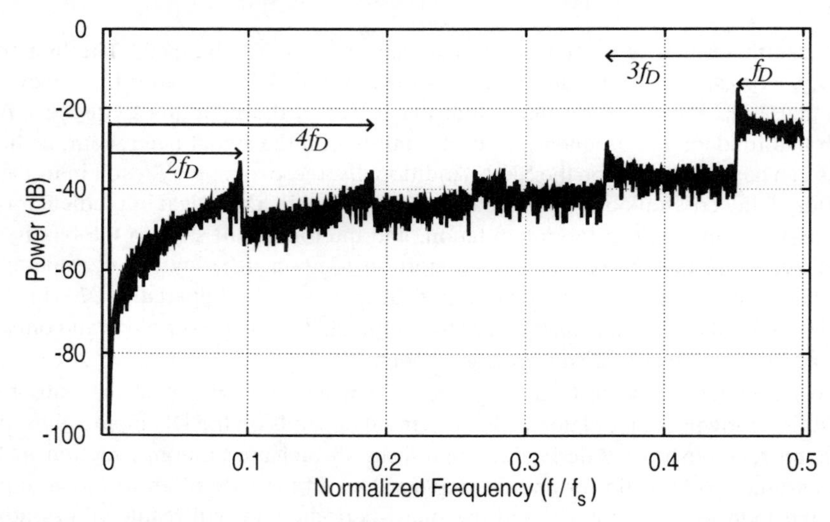

Figure 4.6: *Spectrum of a first order lowpass modulator with sine wave input (simulation, 2K bins).*

The low frequency distortion components $f_{He,n}$ in the output of the modulator cause a deviation in the SNR vs. input power characteristic of the first order SDM. Caused by the amplitude distribution of a sine wave, most of the distortion power will be located near the frequencies $n \cdot f_D$. For large input amplitudes, these distortion components will not fall inside the signal bandwidth $[0, f_b]$ in which the SNR is measured. The quantization noise within this bandwidth will contain few spurious tones and the SNR of the modulator will be near the value predicted by the linear model. When the input amplitude decreases, a larger fraction of the power of these distortion components will shift into the signal bandwidth, causing an additional decrease in the SNR. Note that the perceptual performance will decrease even more, due to the correlated in-band distortion (tones). When the amplitude is decreased below a threshold, the lowest frequency distortion component $2f_D$ (see Fig. 4.6) falls inside the signal bandwidth. As a result, the total quantization noise power will strongly depend on the deterministic distortion components. The power of these distortion components depends on the power of the input signal, and the SNR will reduce at a lower rate than predicted by the linear model. This causes a change in the slope of SNR vs. input power characteristic. In Fig. 4.7 this characteristic is shown for two oversampling ratios. The input power is determined relative to the quantizer output power of $q^2/4$. The actual position of the

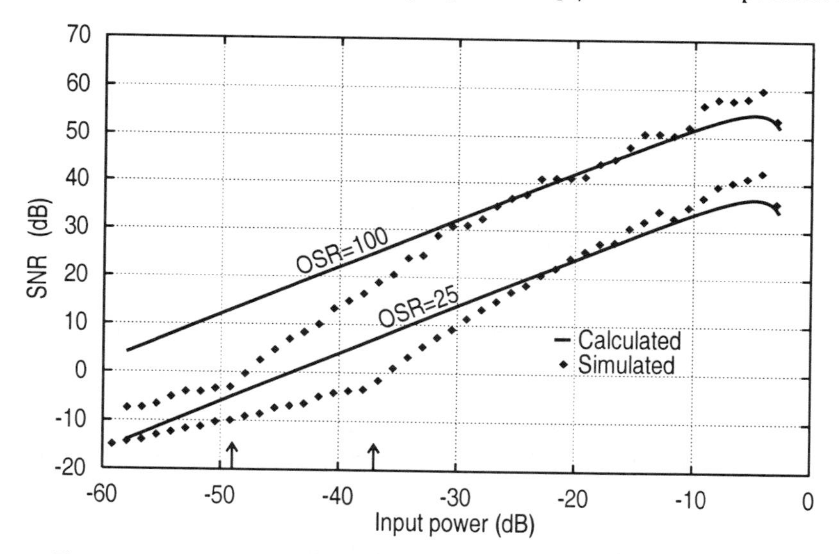

Figure 4.7: Simulated and predicted SNR of a first order lowpass SDM.

threshold depends on the OSR and can be calculated as follows. The slope of the SNR curve will change when the lowest frequency distortion component $2f_D$ is equal to the signal bandwidth f_b. Substituting (4.28) and $f_b = f_s/(2 \cdot \mathrm{OSR})$ gives

$$2\frac{A_t}{q}f_s = \frac{f_s}{2 \cdot \mathrm{OSR}} \tag{4.29}$$

from which the threshold amplitude can be calculated:

$$A_t = \frac{q}{4 \cdot \mathrm{OSR}} \tag{4.30}$$

For an OSR of 100, the threshold amplitude equals $A_t = q/400$, giving a sine wave power of -49dB compared the quantizer output power. In the case that the OSR=25, the threshold is located at -37dB.

Second and Higher Order Modulators

Exact analysis of the behavior of tones in second and higher order modulators is difficult. Even when exact expressions can be derived (see [37]), they do not always give a clear insight in the effects on, for example, the behavior of in-band noise as a function of the input signal. Although not supported by a rigorous mathematical foundation, a rough explanation of the tonal behavior of higher order modulators will be given hereafter, based on the results for a first order modulator, exact analysis found in literature and empirical results.

A second order lowpass modulator with an $\mathrm{NTF}(z) = (1 - z^{-1})^2$ has a loop filter $G(z)$ which can be expressed in a single and double integrator term:

$$G(z) = \frac{2z^{-1} - z^{-2}}{(1 - z^{-1})^2} = \frac{z^{-2}}{(1 - z^{-1})^2} + \frac{2z^{-1}}{1 - z^{-1}} \tag{4.31}$$

When an input signal is small enough, it will fall within the dead zone of a first order modulator. The loop filter output resulting from the right hand term in (4.31) will not be large enough to disturb the idle pattern. As a result, the (double integrating) left hand term will be responsible for disturbing the idle pattern. In a second order modulator, the idle pattern results in a "1100" sequence in the output of the modulator, corresponding to an idle "frequency" of $f_s/4$. Disturbing the idle pattern by applying an input will cause tones near the idle frequency that are effectively modulated by the input signal. The behavior of these tones and their effect on the SNR performance of the modulator will be investigated later. First, the behavior for large input amplitudes will be discussed.

In the case that the amplitude of the input signal exceeds the dead-zone threshold value, the right hand term in (4.31) will so large that the output pattern of the second order modulator is disturbed. The right hand term represents a single integrator which has only one delay (z^{-1}). The output of the modulator can be affected within one sample period. This suggests that for large input amplitudes, the output of the second order modulator will contain tones near $f_s/2$ which will behave similarly as in the case of the first order modulator. Simulations of this second order modulator indeed shows tones near $f_s/2$ which are frequency modulated by the input. Figure 4.8 shows an example of the output spectrum of the modulator with a sine wave input signal. The bandwidth of the frequency modulated distortion components near $f_s/2$ can be calculated using (4.28). This first-order behavior of the tones can also be found in third and higher order modulators, albeit less pronounced as in the first and second order SDMs.

Now, the tonal behavior for small input signals will be discussed. In the case of a very small DC input value at the input of the second order modulator, the double integrating term in (4.31) would result in a quadratic curve in the output of the loop filter. A similar signal could be obtained in a first order modulator by first integrating the input signal before applying it to the modulator. This would suggest that for very small input signals, the tones in a second order modulator resemble the tones of a first order modulator with an integrator preceding the actual input of the modulator. As a result, the maximum deviation or bandwidth of the distortion components would not

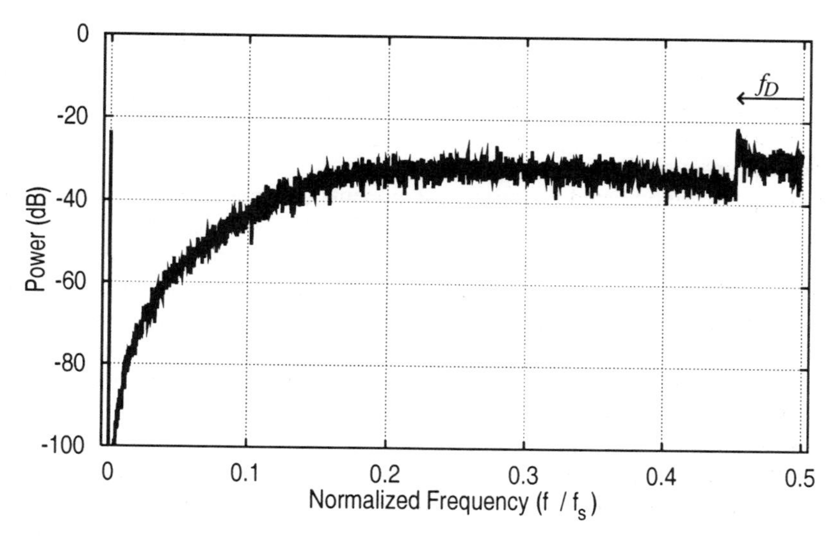

Figure 4.8: Spectrum of a second order lowpass modulator with large amplitude sine wave input (simulation, 2K bins).

only depend on the amplitude, but also on the frequency of the input signal. In the case of an input sine wave $A_i \sin(2\pi f_i t)$, the bandwidth of the distortion components would be a multiple of f_{D2}, with

$$f_{D2} = \frac{A_i}{2\pi f_i q} f_s \tag{4.32}$$

As an example, Fig. 4.9 shows a typical output spectrum of the second order modulator with a very small input amplitude sine wave. Similar to Fig. 4.6, the distortion

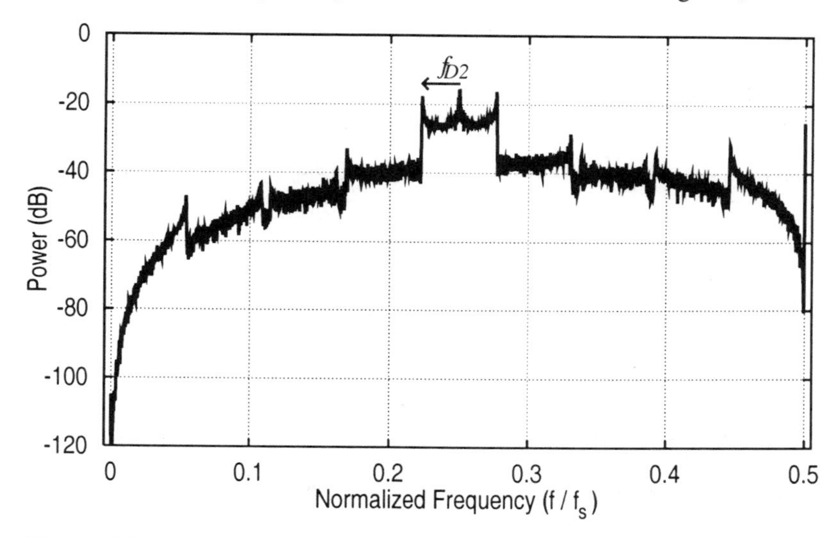

Figure 4.9: Spectrum of a second order lowpass modulator with small amplitude sine wave input (simulation, 2K bins).

components are frequency modulated by the input signal. In Fig. 4.10 the relationship between the maximum deviation f_{D2} (see Fig. 4.9) of the distortion components, and the input amplitude A_i and the input frequency f_i is shown. The bandwidth of the modulated tones around $fs/4$ indeed depends proportionally to the input amplitude and inverse proportionally to the input frequency.

The frequency modulated distortion components or tones located around $f_s/4$ at low input amplitudes have a noticeable effect on the SNR of the modulator. When the input frequency and amplitude are such that the bandwidth of these distortion components is small, they will be located near $f_s/4$. In the case that the amplitude is increased or the input frequency is decreased, the bandwidth of the distortion components will increase and the distortion components with the lowest frequency $f_s/4 - f_{D2}$ will move inside the signal bandwidth $[0, f_b]$. As most of the power of the distortion components is located near $f_s/4 \pm f_{D2}$ (see Fig. 4.9), a sudden decrease in the SNR can then be expected. When measuring the SNR vs. input power characteristic, a noticeable drop in the SNR can be found at relatively low input amplitudes (see Fig. 4.11). The exact position of this sudden decrease depends on the signal bandwidth (or OSR) and the input signal frequency with which the SNR characteristic is measured and can be calculated using (4.32). The sudden drop in SNR will occur at an input amplitude A_{t2} for which

$$f_b = \frac{f_s}{4} - f_{D2} \tag{4.33}$$

Substituting (4.32) in (4.33) and using $f_b = f_s/(2 \cdot \mathrm{OSR})$ gives

$$\frac{f_s}{2 \cdot \mathrm{OSR}} = f_s \left(\frac{1}{4} - \frac{A_{t2}}{2\pi f_i q} \right) \tag{4.34}$$

from which the relative input amplitude can be calculated:

$$A_{t2} = 2\pi q f_i \left(\frac{1}{4} - \frac{1}{2 \cdot \mathrm{OSR}} \right) \tag{4.35}$$

The relative frequency f_i/f_s of the input signal with which the SNR was measured in Fig. 4.11 was equal to $f_i/f_s = 6.37 \cdot 10^{-4}$. With an OSR = 100, the sudden drop

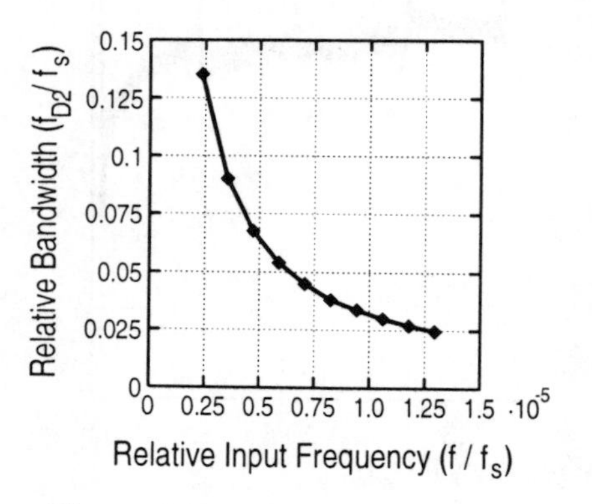

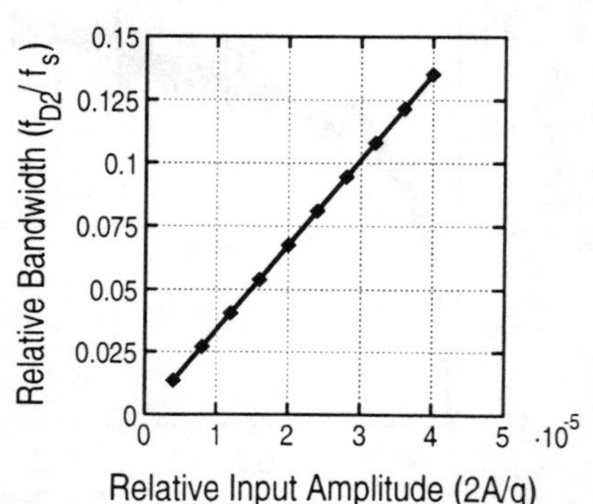

Figure 4.10: *Bandwidth of the distortion components of a second order modulator with a small amplitude sine wave as a function of input frequency (left) and input amplitude (right).*

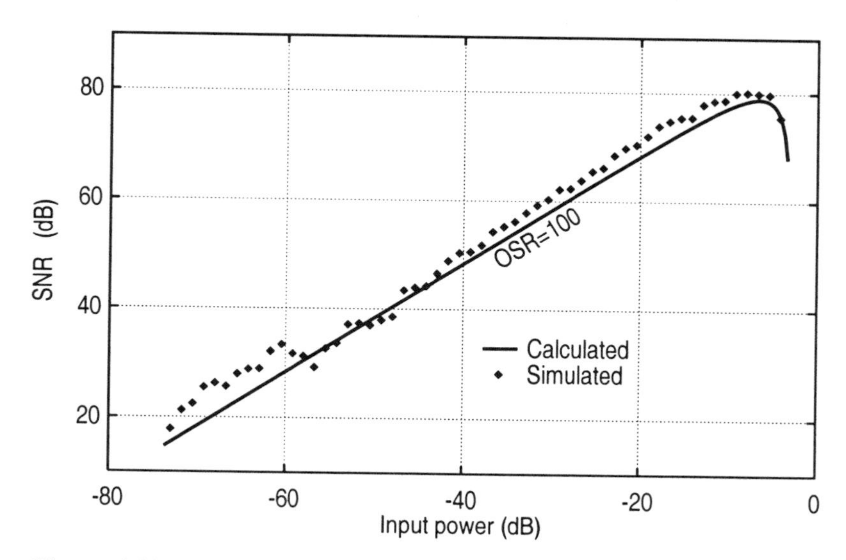

Figure 4.11: Simulated and predicted SNR of a second order lowpass SDM.

in the SNR is located at -57dB relative to the quantizer output power $q^2/4$. Because this "second-order" behavior of the tones can also be found in third and higher order modulators, the noticeable drop in SNR will also be present in the corresponding SNR vs. input power characteristics (see for example [2], p. 147). In Fig. 4.12 this characteristic is shown for a third order lowpass modulator, together with the SNR values according to the linear model. The SNR of the third order SDM was measured with the same input frequency of $f_i/f_s = 6.37 \cdot 10^{-4}$. Therefore, the drop in SNR is located at the same input amplitude as in the case of the second order modulator.

4.3 Dither and Chaotic Modulators

The effects described in the previous section cause a reduction in the measured performance of the modulator. Moreover, they also seriously degrade the perceptual performance. Spurious tones and modulated noise in particular are highly undesired artifacts as they can be perceived even when exceeded (in power) by the in-band noise of the system. This tonal behavior of the SDM can be reduced or even eliminated by applying a dither signal and/or by making the modulator chaotic. These solutions also reduce related artifacts such as sudden drops in the SNR vs. input power characteristic. However, these improvements always result in an overall SNR degradation.

4.3.1 Dither

As was explained in the previous section, the spurious tones are the result of the sampled nature of the modulator and the nonlinear transfer of the quantizer. Even with a large amplitude signal applied to the modulator, the quantizer input signal will be

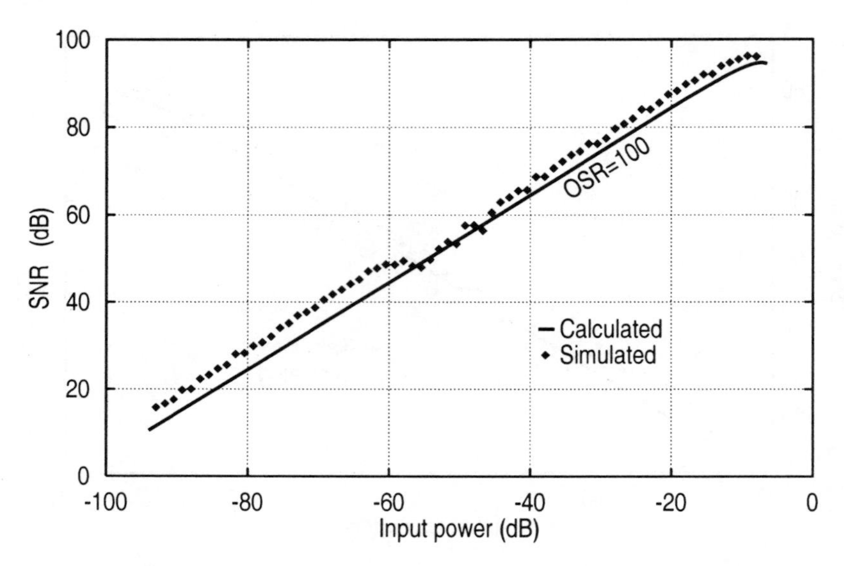

Figure 4.12: Simulated and predicted SNR of a third order lowpass SDM.

highly correlated with the quantizer output signal due to the feedback. In order to decorrelate the quantizer input signal, a dither signal can be added as depicted in Fig. 4.13 [6]. The dither signal effectively disturbs the coding patterns in the output of

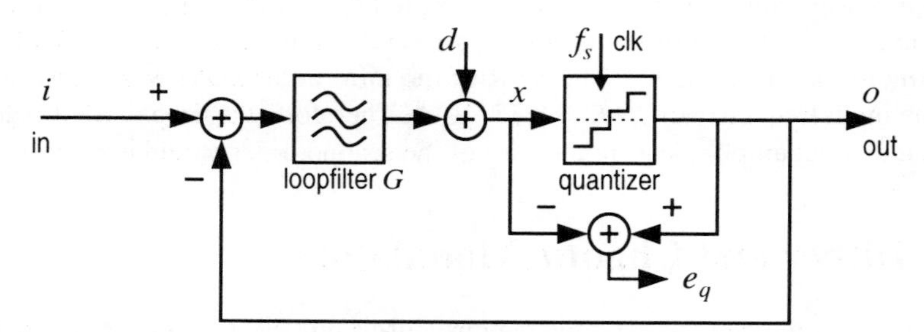

Figure 4.13: An SDM with a dither signal added to the quantizer input.

the modulator in a random fashion. When the dither signal is large enough, the output of the quantizer will be perceptually free of tones. The exact amount of dither that has to be added to the input of the quantizer is still subject of discussion ([2], chapter 3), but is commonly set to have a peak value of half the quantizer step size q.

Adding dither to the quantizer input reduces or even eliminates spurious tones and modulated noise, but also reduces the SNR of the modulator. As the dither is added at the quantizer input it is also noise shaped. The output signal of the dithered SDM in Fig. 4.13 can be written as:

$$O(z) = \frac{G(z)}{1+G(z)}I(z) + \frac{1}{1+G(z)}E_q(z) + \frac{1}{1+G(z)}D(z) \tag{4.36}$$

in which $D(z)$ is the z-transform of the dither signal d. When the quantizer error signal e_q and the dither signal d are assumed to be uncorrelated, a linear prediction model similar to Fig. 4.2 can be used to predict the in-band noise in the output of the quantizer.

4.3.2 Chaotic Modulators

Another method to reduce the pattern noise and tones in the output of a modulator is to design the loop filter such that the SDM exhibits chaotic behavior [39]. In the context of dynamical systems, chaotic behavior or chaos is defined as follows [40]:

> **The behavior of a system is called *chaotic* when a very small perturbation in the initial state of the system results in asymptotically diverging but bounded state space trajectories of the system.**

The behavior of a chaotic system is not steady state, periodic or quasi-periodic. An SDM can be made chaotic by placing the poles of the loop filter $G(z)$ outside the unit circle [41]. Even though the loop filter then becomes unstable in a linear sense, the overall modulator can still be stable[1] due to the nonlinear feedback. The unstable loop filter causes exponentially diverging trajectories which are kept within certain bounds by the nonlinear feedback. As a result, the coding patterns which cause spurious tones in the output of the modulator are disturbed. The quantization error spectrum resulting from a chaotic modulator much resembles that of a noise shaped white random signal.

As in the case of dither, reducing the tones in the output by making the SDM chaotic results in a decrease of the SNR. By placing the poles of the loop filter outside the unit circle instead of on the unit circle, the in-band gain of the NTF is increased. The increase of the in-band noise decreases the SNR of the modulator.

4.4 Non-Ideal Implementation

The theoretically achievable performance of an SDM depends on the order of the loop filter, the resolution of the quantizer and the oversampling ratio. The actual performance also depends on the effects of non-ideal implementation. These effects include a limited gain of the loop filter, noise introduced by the implemented circuitry, crosstalk from clock signals and distortion caused by the feedback DAC.

4.4.1 Limited gain

Caused by the limited gain of the actual loop filter implementation, the poles of the loop filter transfer function cannot be placed exactly on the unit circle. This causes an increase in the in-band gain of the NTF and thus an increase in the idle channel noise. As a result, the DR and SNR of the modulator will decrease. The effects of a limited DC gain of the integrators on the behavior of lowpass modulators has been

[1] For definition of stability in the context of sigma delta modulators see chapter 5.

examined in detail in [42]. However, the loss of DR is not addressed. An estimate of the degradation in the DR can be obtained using the linear prediction of section 4.1. Consider the class of lowpass modulators with loop filter

$$G(z) = \frac{(1-az^{-1})^N}{(1-bz^{-1})^N} - 1 \tag{4.37}$$

A limited gain of the integrators used to implement this transfer function can be modeled by a reduction of the parameter b. When the integrators have a DC gain G_{DC} the parameter b can be determined by

$$b = e^{-\frac{1}{G_{\mathrm{DC}}-1}} \approx 1 - \frac{1}{G_{\mathrm{DC}}} \tag{4.38}$$

Provided that the maximum input signal amplitude does not change, the loss of DR is equal to the increase of the idle channel noise resulting from a reduction in b. The in-band idle channel noise of a lowpass modulator with loop filter (4.37) is calculated by:

$$\begin{aligned} N_{o,idle} &= \frac{2q^2}{24\pi}\int_0^{\theta_b}\left|\frac{1}{1+c_g G(e^{j\theta})}\right|^2 d\theta = \\ &\frac{q^2}{12\pi}\int_0^{\theta_b}\left|\frac{(1-be^{-j\theta})^N}{(1-c_g)(1-be^{-j\theta})^N + c_g(1-ae^{-j\theta})^N}\right|^2 d\theta \end{aligned} \tag{4.39}$$

in which c_g is determined by solving eq. (4.5). Figure 4.14 shows the reduction in DR compared to the ideal case of $b = 1$ for SDMs with the loop filter parameters of Table 4.1.

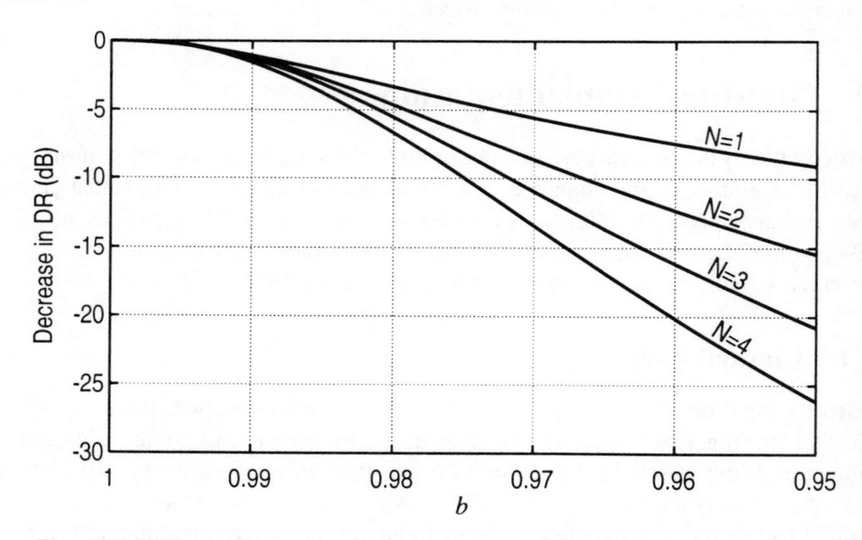

Figure 4.14: *Reduction of the DR as a function of b relative to the DR at b = 1 of a lowpass modulator of order N (OSR=100).*

Similarly, the reduction in SNR can be calculated for a limited quality factor Q of the loop filter in bandpass modulators. For a second order discrete time resonator section defined by

$$G(z) = \frac{z^{-1}}{(1 - be^{j\theta_0}z^{-1})(1 - be^{-j\theta_0}z^{-1})} \tag{4.40}$$

the relationship between the quality factor Q and the parameter b can be determined. The quality factor Q is defined (see [43]) as the ratio between the average energy U stored in the loop filter and the dissipation per radian $-dU/d\phi$:

$$Q = \frac{U}{-dU/d\phi} \tag{4.41}$$

For a continuous time filter tuned at ω_0, this equation can be rewritten as

$$-\frac{U}{Q} = \frac{dU}{dt} \cdot \frac{dt}{d\phi} = \frac{dU}{dt} \cdot \frac{1}{\omega_0} \tag{4.42}$$

Solving the energy U from this differential equation gives:

$$U(t) = U_0 e^{-\frac{\omega_0}{Q}t} \tag{4.43}$$

In the case of a second order continuous time filter section the quality factor Q corresponds to the ratio between the center frequency f_0 and the so-called -3dB bandwidth $\mathrm{BW}_{-3\mathrm{dB}}$. This bandwidth is defined as the distance between the frequencies $f_{-3\mathrm{dB}}$ at which the gain of the filter is -3dB below the maximum value at f_0. The energy stored in the discrete time resonator cf. (4.40) can be approximated by:

$$U(kT_s) \approx b^{2k} \cdot U_0 \tag{4.44}$$

This expression can be rewritten using an exponential form:

$$U(kT_s) \approx U_0 \cdot e^{2k \cdot \ln(b)} \tag{4.45}$$

Substituting $k = t/T_s$ and comparing (4.45) and (4.43) results in

$$Q \approx -\frac{\theta_0}{2 \cdot \ln(b)} \qquad \text{or:} \qquad b \approx e^{-\frac{\theta_0}{2Q}} \tag{4.46}$$

This relationship between Q and b can also be derived by the continuous time to discrete time transformation of a continuous time resonator with quality factor Q (see sec. 6.4). The curves shown in Fig. 4.14 will exhibit an exponential form when plotted as a function of Q. As in the case of the lowpass modulators, the loss of DR for a set of bandpass SDMs with loop filter

$$G(z) = \frac{(1 - ae^{j\theta_0}z^{-1})^{N/2}(1 - ae^{-j\theta_0}z^{-1})^{N/2}}{(1 - be^{j\theta_0}z^{-1})^{N/2}(1 - be^{-j\theta_0}z^{-1})^{N/2}} - 1 \tag{4.47}$$

is calculated by determining the increase of the idle channel noise:

$$N_{o,idle} = \frac{2q^2}{24\pi}\int_0^{\theta_b}\left|\frac{1}{1+c_g G(e^{j\theta})}\right|^2 d\theta \tag{4.48}$$

The decrease in DR for bandpass modulators tuned at $\theta_0 = \pi/2$ (or $f_s/4$) as a function of Q is shown in Fig. 4.15. The parameter a is set to the minimum required value for stability (see sec. 5.6.5).

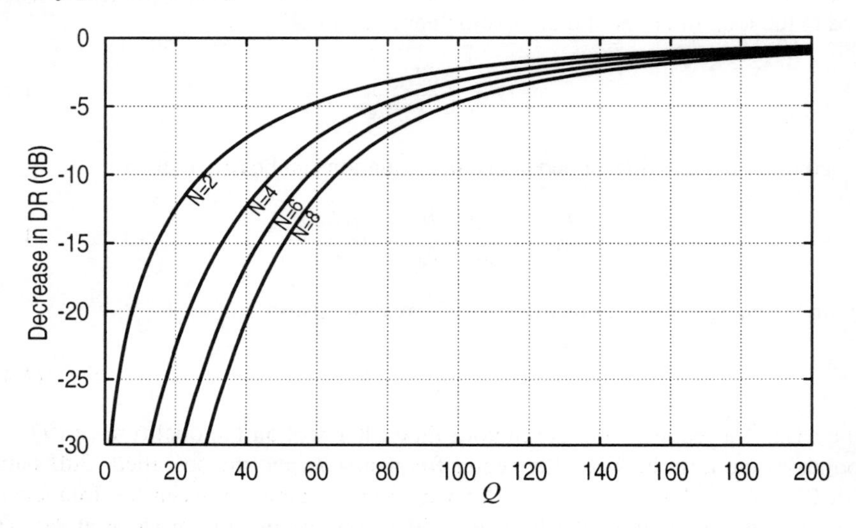

Figure 4.15: Reduction of the DR as a function of Q relative to the DR at $Q = \infty$ of a bandpass modulator of order N tuned at $f_s/4$ (OSR=100).

4.4.2 Noise

In section 4.1 an estimate for the performance of an ideal SDM was found, under the assumption that the quantization errors could be considered as additive white noise. In a practical implementation of an SDM, the circuitry will introduce additional noise in the form of thermal noise, 1/f-noise, etc. which is generally independent of the input signal. The circuit noise can be accounted for by a single noise source which adds to the input signal. This noise will not be shaped by the SDM, and adds directly to the quantization noise in the output of the SDM. The equivalent amount of noise in the input of the modulator can easily be calculated. In Fig. 4.16 a continuous time SDM is shown with several noise sources at different positions inside the SDM loop. The noise n_{fil} introduced by the filter is suppressed by the feedback loop. The noise can be expressed as an equivalent input noise by dividing it by the loop filter transfer $G(p)$. The noise n_{DAC} introduced by the DAC adds directly to the input of the SDM. The noise n_{ADC} at the output of ADC (or quantizer) does not have a significant effect on the performance of the modulator and can be neglected. This is due to the fact that the output signal o is a digital signal. Both the DAC inside the SDM loop and the system connected to the output of the SDM will have large noise margins. When the

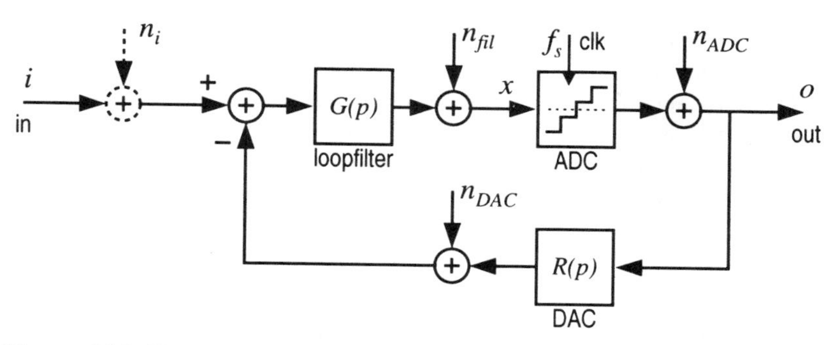

Figure 4.16: *SDM with several noise sources and equivalent input noise source (dashed).*

noise does not exceed these margins it will have no effect on the performance of the SDM. By assuming that all the noise sources are spectrally white and independent, the equivalent input noise power N_i can be calculated by:

$$N_i = \frac{N_{\text{fil}}}{\int_{\omega=0}^{\infty} |G(j\omega)|^2} + N_{\text{DAC}} \tag{4.49}$$

The actual implementation determines which of the two noise sources is dominant. Even though the noise contribution of the filter is divided by the filter transfer, it may prove to be dominant in an actual realization.

The circuit noise generally degrades the performance of an SDM, but can have a positive side effect. When the equivalent circuit noise at the input of the SDM is larger than the quantization noise, tones and pattern noise will be masked. Additionally, the circuit noise can act as a dither signal at the input of the quantizer, thus reducing pattern noise and tones in the output of the modulator even more.

4.4.3 Crosstalk and Distortion

Crosstalk and distortion can be analyzed in much the same way as circuit noise. Nonlinearities in the loop filter and DAC introduce distortion signals such as harmonic and intermodulation products. Crosstalk, e.g. from the output to the input of the quantizer, can also introduce signal dependent distortion. These distortion signals can be modeled by adding signals at several locations in the modulator loop similar to the noise in Fig. 4.16. As in the case of the circuit noise, the distortion introduced by the loop filter is suppressed by the feedback loop. The distortion of the DAC is not suppressed and adds directly to the input signal.

As was mentioned in section 3.6.5, the distortion in a one-bit DAC is not caused by level mismatch. Level mismatch in a one-bit quantizer only results in a DC offset and amplitude change which are linear deviations. Distortion in a one-bit DAC can be caused by inter-symbol interference. The DAC generates a positive or negative pulse, depending on the input signal. Often a so-called Non Return-to-Zero (NRZ) pulse is used to convert the discrete time input signal into a continuous time signal. The duration of an NRZ pulse is equal to the sampling period T_s. Because these pulses in practice will have a limited rise and fall time, the energy of a pulse will depend on whether or not the output level will have to change value. The energy of a positive pulse ('1') following a negative pulse will be less than when preceded by a positive pulse. In Fig. 4.17 the output of a one-bit quantizer with limited rise and fall times is shown. The errors caused by the absence of rising and falling edges are indicated by gray areas. Because the distortion depends on the preceding symbol, it is signal dependent and will result in harmonic distortion components and intermodulation products.

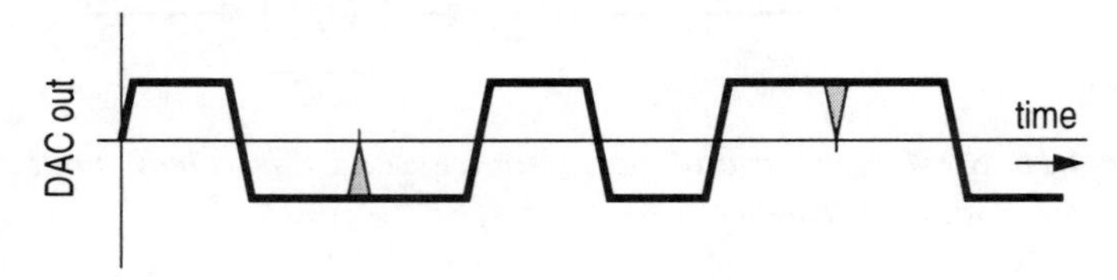

Figure 4.17: *Distortion due to limited rise and fall times of a DAC using NRZ pulses.*

A solution for the reduction of the effects of limited rise and fall times is the use of Return-to-Zero (RTZ) pulses. The duration of these pulses is less than the sample period T_s. As a result, every single pulse at the output of the DAC will have a rising and falling edge and the energy of each pulse will be (almost) identical. An example of the output signal of a DAC using RTZ pulses is shown in Fig. 4.18.

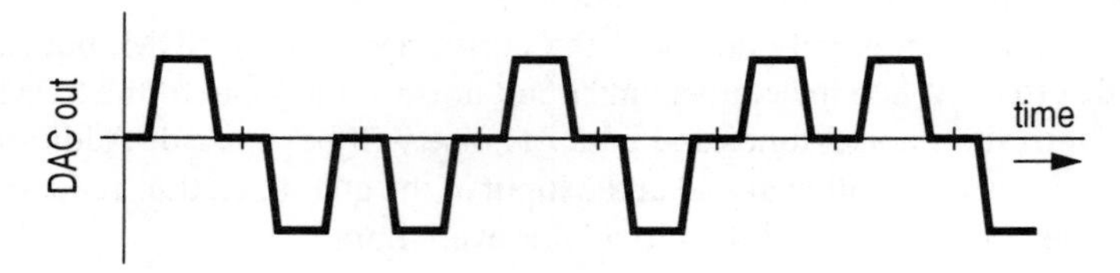

Figure 4.18: *Output signal of a DAC using RTZ pulses.*

4.5 Conclusions

The Signal-to-Noise performance of an SDM can be estimated by using a linear prediction model. Modeling a one-bit quantizer by a gain and the addition of white noise results in a good estimate of the SNR as a function of the input amplitude. Because of the linear modeling, this method fails to predict the effects of idle patterns and tones on the performance of the modulator. Idle patterns cause a dead-zone in which the output does not contain a quantized representation of the input signal. The deterministic behavior of the SDM causes tones in the output of the modulator. For lowpass modulators, these tones depend on the DC value of the input signal. For slowly varying input signals, the tones can be viewed as frequency modulated distortion components. By analyzing the behavior of these tones, several deviations in the actual SNR performance can be explained. In-band tones cause bumps and slope changes in the SNR vs. input power characteristics of SDMs.

CHAPTER 5

STABILITY

The high Signal-to-Noise ratio that can be achieved by high-order one-bit sigma delta modulators has led to the wide spread application of these modulators as oversampled A/D converters. However, the stability of the nonlinear feedback loop of the modulators has yet to be proven. Caused by the nonlinearity of the quantizer, stability methods for linear systems such as the Routh-Hurwitz test are not applicable. The nature of the instabilities also requires a refinement of the definition of stability within the context of SDMs.

5.1 Definitions

Consider the following discrete time nonlinear system [44, 45]. The state of the system is represented by the state vector $\mathbf{x}$[k], in which $k = 0, 1, \ldots$ represents the sample number. The autonomous behavior of the system in time is described by:

$$\mathbf{x}[k+1] = f(\mathbf{x}[k]) \qquad \text{with:} \qquad \mathbf{x}[0] = \mathbf{x}_0 \tag{5.1}$$

in which f is generally a nonlinear time-invariant function, and $\mathbf{x}_0$ is the initial state of the system. For DC input conditions, the discrete time sigma delta modulator shown in Fig. 5.1 can be written in the form of (5.1), with $\mathbf{x}$ corresponding to the states of the loop filter $G(z)$. The stability of the system described by (5.1) is defined by:

> **An autonomous system is called stable when for a certain bounded set of initial states $\mathbf{x}_0 \in \mathbf{C}$, the state $\mathbf{x}[k]$ of the system will be bounded for $k \to \infty$. The system is called globally stable when the system is stable for any bounded initial state $\mathbf{x}_0$.**

A state $\mathbf{x}[k]$ is called bounded when $||\mathbf{x}[k]|| < \infty$ in which $||.||$ represents the Euclidean norm. As will be shown, this definition of stability is not very suitable in the context of sigma delta modulators. A more practical definition of stability for SDMs will therefore be introduced later.

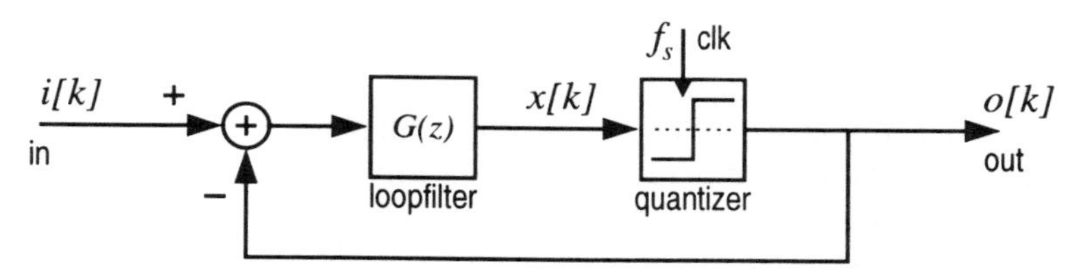

Figure 5.1: *A single loop one-bit discrete time sigma delta modulator.*

The behavior of the system, or in other words: the trajectories of the solutions of the system through state space, is characterized by special points in state space. Similar to linear systems, the nonlinear system can have equilibrium points $\mathbf{x}^\star$ for which:

$$f(\mathbf{x}^\star) = \mathbf{x}^\star \tag{5.2}$$

Consequently, when $\mathbf{x}[k_0] = \mathbf{x}^\star$ then $\mathbf{x}[k] = \mathbf{x}^\star$ for all $k \geq k_0$. Concerning the stability of such an equilibrium point (e.p.) several remarks can be made [45, p. 165]. The equilibrium point $\mathbf{x}^\star$ is called

1. *Stable* if for any given radius $\varepsilon > 0$ and $k_0 \geq 0$ there exists a neighborhood $\delta = \delta(\varepsilon, k_0)$ such that $||\mathbf{x}[k_0] - \mathbf{x}^\star|| < \delta$ implies $||\mathbf{x}[k] - \mathbf{x}^\star|| < \varepsilon$ for all $k > k_0$. The e.p. is called *uniformly stable* when δ does not depend on k_0; and *unstable* when it is not stable.

2. *Attractive* if there exists a radius $\mu = \mu(k_0)$ such that $||\mathbf{x}[k_0] - \mathbf{x}^\star|| < \mu$ implies $\lim_{k \to \infty} \mathbf{x}[k] = \mathbf{x}^\star$. The e.p. is said to be *uniformly attractive* if μ does not depend on k_0; *repelling* if the e.p. is not attractive.

3. *(Uniformly) asymptotically stable* if it is both (uniformly) attracting and (uniformly) stable. The e.p. is said to be *globally (uniformly) asymptotically stable* when the radius of the attracting region $\mu = \infty$.

In addition to equilibrium points, nonlinear systems can exhibit limit cycles. A limit cycle in a discrete time system is a set of points $\mathbf{x}_1, \mathbf{x}_2, \ldots, \mathbf{x}_M$ for which holds that $\mathbf{x}_{i+1} = f(\mathbf{x}_i)$ for $1 \leq i \leq M-1$, and $f(\mathbf{x}_M) = \mathbf{x}_1$. A limit cycle represents an oscillatory motion of the state of the system which is periodic with period M as $f^M(\mathbf{x}_1) = \mathbf{x}_1$. Note that in this respect, equilibrium points can be regarded as limit cycles of period $M = 1$. Similar to equilibrium points, a limit cycle can be attracting or repelling and stable or unstable.

Equilibrium points and limit cycles are a special case of a positively invariant limit set [45, p. 209]. A positively invariant set in general is defined as a set $\mathbf{C}$ for which holds that

$$\mathbf{x} \in \mathbf{C} \Rightarrow f(\mathbf{x}) \in \mathbf{C} \tag{5.3}$$

A positively invariant limit set $\Omega(\mathbf{x}_0)$ is defined by:

$$\Omega(\mathbf{x}_0) = \{\mathbf{y} \mid \mathbf{x}[k_i] \to \mathbf{y} \text{ as } k_i \to \infty \text{ for some subsequence } \{k_i\} \text{ of } \mathbb{Z}^+\} \tag{5.4}$$

Roughly said, the positively invariant limit set is equal to the set of points to which the state $\mathbf{x}$ of the system will converge for some $\mathbf{x}_0$ and $k \to \infty$. Note that such set can again be attracting or repelling and stable or unstable. In practical systems, unstable limit sets will not be observed, as any perturbation will move the system away from the unstable e.p. , limit cycle or limit set. In order to verify the existence of these unstable e.p.'s and limit cycles, a reverse time simulation can be done. In eq. (5.1) the time parameter k is then replaced by $-k$, thus changing stable e.p.'s and limit cycles into unstable ones and vice versa.

A positively invariant limit set can result in an oscillatory motion of the state of the system which is quasi-periodic or even a-periodic. For example, consider the following nonlinear system:

$$\theta[k+1] = 2\theta[k] \mod 2\pi \quad (5.5)$$

$$r[k+1] = \sqrt{r[k]} \quad (5.6)$$

in which $\mathbf{x} = (r, \theta)$ with $r \geq 0$ and $\theta \in [0, 2\pi]$. This system has two equilibrium points $(0,0)$ and $(1,0)$. For $0 < r_0 < 1$, $\lim_{k \to \infty} r[k] = 1$. The solutions will spiral towards the unit circle, away from the origin: the e.p. $(0,0)$ is unstable. For $r_0 > 1$, $\lim_{k \to \infty} r[k] = 1$ the solutions will also spiral towards the unit circle: the e.p. $(1,0)$ is stable. When $r_0 = 1$, the solutions will cycle on the unit circle. However, these solutions will be periodic if and only if $\theta[k] = 2k\pi/(2^m - 1)$ for some integers $k, m \geq 0$. This set of periodic solutions is called a limit cycle. For other values of θ, the unit circle is a positively invariant limit set, but not a periodic limit cycle.

For reasons of simplicity, a positively invariant limit set resulting in quasi-periodic behavior will also be referred to as a limit cycle. In this context, the behavior is called quasi-periodic if the trajectory of the state can be described by:

$$\mathbf{x}[k] = \mathbf{g}(kT_s; T) \quad \text{with:} \quad T/T_s \in \mathbb{R} \setminus \mathbb{Q} \quad (5.7)$$

in which $\mathbf{g}$ is a periodic continuous time function with period T.

In [46] LYAPUNOV showed that a sufficient condition for the autonomous system to be globally stable and for the state to remain bounded, is the presence of a globally asymptotically stable equilibrium point. This condition can be extended by requiring the presence of a globally asymptotically stable limit set.

Unfortunately, the definition of stability and the corresponding sufficient condition defined by LYAPUNOV are not suitable for practical sigma delta modulators. First of all, the sufficient condition was stated for an autonomous system. Although a similar condition can be stated for a special class of input signals ("bounded squared-integrable signals"), the actual input signals applied to SDMs, e.g. sinusoidal signals, do not belong to this class. Secondly, the boundedness of the states of the sigma delta modulator does not imply correct operation. Firstly, in practical implementations, the states of the SDM should remain within specified boundaries in order to avoid overflow and/or clipping. For example, when the states of the filter are represented by voltages, the positive and negative supply voltage pose an absolute limit to the state values. Note that this implies that practical modulators are always stable in *bounded input bounded state (BIBS)* sense. Secondly, in general the performance of the modulator will deteriorate when the state values become very large compared to the modulator input and output values. This can be made clear by the following observation. In the desired operation of the SDM, the output $o[k]$ of the modulator should represent a (filtered) representation of the input signal $i[k]$ plus distortion components. This means that the input of the quantizer $x[k]$ should be strongly influenced by the input of the modulator. Now assume that the modulator is attracted to a stable limit cycle[1]. When the amplitude of the limit cycle is small, the input signal applied to the modulator can easily

[1] Modulators with a stable loop filter, in practice, are stable in BIBS sense and exhibit stable limit cycles.

change the input of the quantizer. The limit cycle is disturbed, and the output of the modulator will contain a (filtered) representation of the input signal. In the case that the amplitude of the limit cycle is very large (the state values are very large), the input signal applied to the modulator will not disturb the limit cycle significantly. The output of the modulator will mainly be determined by the limit cycle, and not by the input signal. As as result, the performance of the modulator will be low.

The previous observations show that the concept of stability with respect to SDMs and other error feedback coders needs to be redefined. In [47] HEIN and ZAKHOR introduced the term K-stability. A system (or SDM in particular) is called K-stable when, for a given class of input signals, the states of the system are bounded in absolute value by K. As indicated by the authors, K should be normalized to eliminate the effects of so-called equivalent scaling. For example, scaling the filter input signal in Fig. 5.1 with a factor of c does not change the behavior of the one-bit modulator, but does scale the values of the filter states. Unfortunately, a uniform method to normalize the system and have an unambiguous stability criterion is hard to find. Therefore, the concept of stability with respect to SDMs used here will be given using qualitative measures instead of quantitative criteria.

In order to define the concept of stability for SDMs, a distinction will be made between two types of limit cycles based on the amplitude or maximum value of (one of) the states of the loop filter. A limit cycle will be called an *idle pattern* when it results in a small amplitude periodic signal at the output of the loop filter. The gain of the filter is usually very large inside the intended signal band; the filter is said to be tuned to the input signal frequency range. As the output signal of the filter should be small in the case of an idle pattern, the frequency[2] of the idle pattern should not be near the signal band. In contrast to an idle pattern, a limit cycle is called a *large signal limit cycle* when it results in a very large amplitude at the output of the loop filter. The frequency of a stable large signal limit cycle will therefore be located near the tuning frequency of the loop filter.

As an example, the idle pattern and a large signal limit cycle of a third order one-bit lowpass modulator are shown in Fig. 5.2. The loop filter of the modulator is given by:

$$G(z) = \frac{(z-0.3)^3}{(z-0.95)^3} - 1 \tag{5.8}$$

Depending on the initial state of the loop filter, the filter state or "solution" of the modulator converges to the idle pattern or the large signal limit cycle (no input signal is applied to the modulator).

Using the distinction between idle patterns and large signal limit cycles an alternative definition of stability with respect to SDMs can be given:

> **A sigma delta modulator will be called stable when for a certain class of input signals, the states of the system are bounded and the modulator is free of large signal limit cycles.**

[2]The frequency of a limit cycle in a one bit SDM is defined as the frequency of the fundamental harmonic of the quantizer output signal when the limit cycle is present.

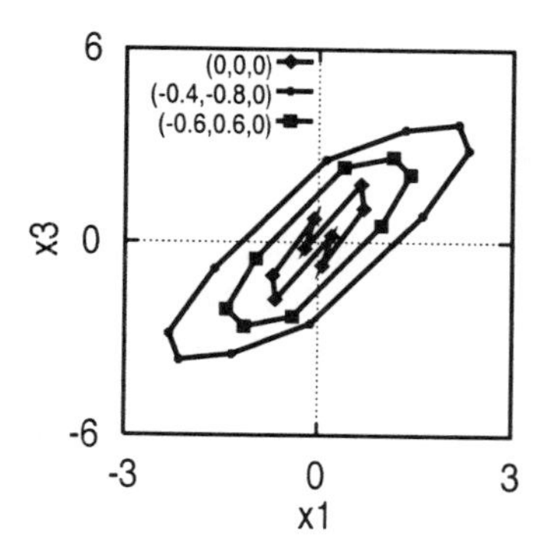

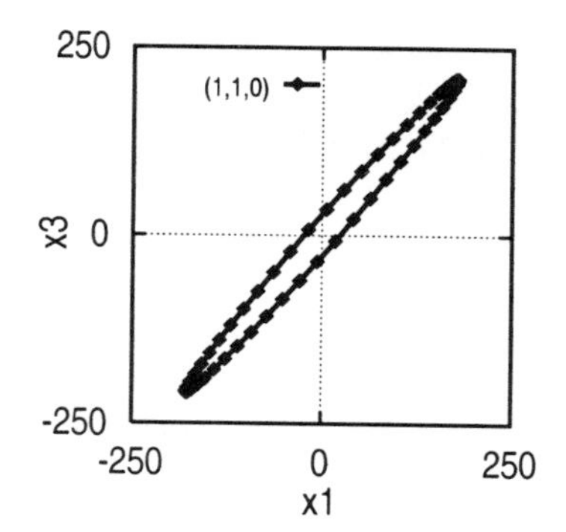

Figure 5.2: Idle patterns (left) and large signal limit cycle (right) of a third order modulator with loop filter (5.8). The coordinates indicate the initial state resulting in the specific limit cycle.

5.2 Stability Analysis Methods and Criteria

As was mentioned in the introduction, stability analysis methods for linear systems cannot be used to examine the behavior of nonlinear systems. Because of the nonlinearity, the superposition principle is not applicable. The response of the system to an input signal cannot be calculated by determining the homogeneous solution and the particular solution. In order to be able to characterize the behavior of nonlinear systems several methods have been developed. In addition, several stability criteria for sigma delta modulators in particular have been introduced.

A) Lyapunov's Method

The direct or second method of LYAPUNOV [46] is a well known method in order to determine the stability of nonlinear systems. The method can be used for proving global stability of equilibrium points and boundedness of states. It is based on finding a so-called Lyapunov function which is positive definite on the neighborhood of an equilibrium point. A real-valued function $V(\mathbf{x})$ is called a Lyapunov function on a set $\mathbf{C}$ when

1. V is continuous on $\mathbf{C}$
2. $\Delta V(\mathbf{x}) = V(f(\mathbf{x})) - V(\mathbf{x}) \leq 0$ whenever $\mathbf{x}$ and $f(\mathbf{x}) \in \mathbf{C}$.

The function $V(\mathbf{x})$ is called positive definite at e.p. $\mathbf{x}^\star$ if

1. $V(\mathbf{x}^\star) = 0$
2. $V(\mathbf{x}) > 0$ for all $\mathbf{x}$ with $||\mathbf{x} - \mathbf{x}^\star|| > \varepsilon$ for some $\varepsilon > 0$.

Using these definitions, an equilibrium point $\mathbf{x}^\star$ is called stable when there exists a positive definite function V on the neighborhood $\mathbf{C}$ of $\mathbf{x}^\star$. Furthermore, the e.p. is asymptotically stable when $\Delta V(\mathbf{x}) < 0$ for all $\mathbf{x}, f(\mathbf{x}) \in \mathbf{C}$, and globally stable when $\mathbf{C}$

is equal to the entire state space and $V(\mathbf{x}) \rightarrow \infty$ as $||\mathbf{x}|| \rightarrow \infty$. The Lyapunov function can be seen as a kind of energy function. Proving that the mapping $f(\mathbf{x})$ will result in an energy decrease for any point in state space ($\Delta V(\mathbf{x}) < 0$) guarantees stability of the system.

Several stability criteria such as the JURY and POPOV frequency domain criteria [44] are derived from Lyapunov's method. Unfortunately, these methods all share some serious drawbacks when applied to sigma delta modulators. Firstly, the method can only be used to ascertain stability by finding a positive definite Lyapunov or energy function. The inability to find such a function does not prove instability of the system. Secondly, the method can only be used to guarantee boundedness of states. As was discussed in the previous section, this is not a useful measure of stability with respect to SDMs.

B) Tsypkin's Method

To circumvent the nonlinear transfer function in feedback systems such as the sigma delta modulator in Fig. 5.1, TSYPKIN introduced the following method to ascertain the existence of limit cycles [48]. The output $o[k]$ of the nonlinearity is assumed to have a certain periodic waveform. The response $x[k]$ of the linear loop filter to the sum of this waveform and the input signal $i[k]$ is then calculated. When the resulting input signal of the quantizer indeed results in the assumed output waveform, the limit cycle is sustained. This method has been applied to SDMs in particular in [47] and [49]. Clearly, in order to determine the overall stability of the system an infinite number of limit cycles should be tested. In order to limit the number of limit cycles to be tested, a priori knowledge of the behavior of the SDM should be available. Hence, this method is not very suitable as a general stability method for sigma delta modulators.

C) Positively Invariant Sets

Another method for investigating the stability of a nonlinear feedback system is the use of positively invariant sets [50]. When a (not necessarily convex) positively invariant set can be found for the solutions of the dynamic behavior of the SDM, the SDM can be considered stable within this set. If the state of the SDM will remain inside this set for a certain class of input signals, the SDM can be considered stable for these signals. Finding an analytical expression for such a positively invariant set is difficult, and current methods rely on numerical solutions. Although this method is highly reliable, the use of extensive computer calculations in the design of an SDM is unattractive.

D) Describing Function Method

The describing function (df) method [44] is a well known method for analyzing the behavior of nonlinear systems. The concept of the df method is to use a linear, but signal dependent approximation of the nonlinear transfer function. This approximation is based on two assumptions:

1. the input signal of the nonlinear element is a sinusoid.

2. the fundamental harmonic of the input signal is prevalent in the output signal of the nonlinear element.

In many practical nonlinear feedback systems these assumptions are validated as the output signal indeed contains a dominant fundamental harmonic. Additionally, the lowpass characteristic of the linear loop filter eliminates the higher harmonics, reducing the input signal of the nonlinear element to a single sinusoid. Limit cycles in such systems often exhibit a sinusoidal waveform.

Even when the assumptions are validated, the effectiveness of the df method depends on the ability to adequately model the nonlinear element. Commonly, the nonlinear element is modeled by a variable gain depending on the magnitude of input sinusoid. This model has also been applied to sigma delta modulators [29, 31, 51]. In section 5.3 it will be shown that despite reasonably accurate results the model fails to predict certain aspects of the SDM behavior. However, the df method does provide an excellent qualitative insight in the dynamic behavior of SDMs, and will be the main method to investigate the stability of SDMs in the rest of this book.

E) Lee's Rule

In addition to general methods for stability analysis of nonlinear systems, several stability criteria or rules of thumb have been introduced for the design of stable sigma delta modulators. A well known rule was introduced by LEE et al. [52]. LEE stated that the out-of-band gain of the noise transfer function should be less than two for zero input stability:

$$|\mathrm{NTF}(z)| < 2 \qquad \text{for:} \quad z = e^{j\theta} \tag{5.9}$$

The basic idea behind this criterion is that the noise transfer function amplifies the out-of-band noise introduced by the quantizer and causes the modulator to become unstable when the amplification is too large. Although this rule of thumb was determined by simulation for a fourth-order lowpass modulator it is now widely used in the design of SDMs [2].

F) Power Gain Rule

Another, but less used experimental rule of thumb for the stability of SDMs was introduced by AGRAWAL and SHENOI [28]. Based on the white noise assumption of quantization errors they suggested that the total power gain of the NTF should be less than three. The power of the quantization errors is assumed to be $q^2/12$. However, the maximum output power of the modulator equals $q^2/4$. As a result, the total power gain of the NTF should not exceed a factor of three to validate the white noise assumption.

In [53] SCHREIER and SNELGROVE presented a comparison of several rules of thumb with experimental stability tests. They showed that both LEE's Rule and the Power Gain Rule are neither necessary nor sufficient to guarantee stability.

5.3 Describing Function Method

In this section the describing function method and its application to stability analysis of SDMs will be investigated more closely. For the analysis the general model shown in Fig. 5.3 will be used. The model contains a loop filter G and a quantizer Q enclosed

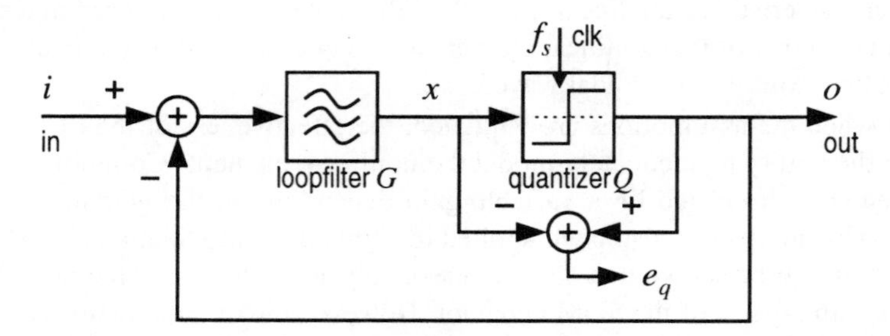

Figure 5.3: *General model of a one-bit SDM for stability analysis using the describing function method.*

by a negative feedback loop. The quantizer is sampled at a rate of f_s samples per second. Loop filter G can either be a discrete time or a continuous time filter. The stability analysis of the loop can be done in the discrete time domain, as the signal path within the loop contains a sampled element (quantizer). A continuous time loop filter can be replaced by an equivalent discrete time filter (see sec. 3.6.4).

As to the df method, the input signal x of the nonlinear quantizer is assumed to be a sinusoid:

$$x[k] = A_x \sin(2\pi f_x k T_s) \tag{5.10}$$

The output signal o of the quantizer can be expressed as a Fourier series with a fundamental frequency equal to the input frequency f_x:

$$o[k] = \sum_{i=0}^{\infty} a_i \sin(2\pi i f_x k T_s + \phi_i) \tag{5.11}$$

The nonlinear transfer of the quantizer Q is now approximated by the response of the fundamental harmonic to the input signal. Generally, this response depends on both the amplitude A_x and the frequency f_x and consists of a gain and phase shift:

$$\tilde{Q}(A_x, f_x) = \lambda(A_x, f_x) \cdot e^{j\phi(A_x, f_x)} \tag{5.12}$$

Commonly, a one-bit quantizer is modeled by a frequency independent global signal gain λ (see [29,31,51]). This gain results from the fixed output amplitude and variable input amplitude of the quantizer. As the output of the one-bit quantizer has a constant amplitude, the gain can vary from $\lambda \to \infty$ when the quantizer input amplitude $A_x = 0$, to $\lambda = 0$ when the amplitude A_x is infinitely large. The linearized transfer of the quantizer can be written in the z-domain as:

$$\tilde{Q}(z) = \lambda \qquad \text{with:} \quad \lambda \in [0, \infty) \tag{5.13}$$

Now that the nonlinear quantizer has been modeled by a linear transfer function, linear stability methods can be applied. A practical method for stability analysis of linear systems is the root locus method. By drawing the locations of the poles of the closed loop system in the complex plane, the stability of the system can be verified. For discrete time systems, the poles should be inside the unit circle in the complex z-plane to guarantee stability. The closed loop signal transfer of the linearized system can be expressed in the appropriate z-domain representatives of input $i]k]$, output $o[k]$:

$$\frac{O(z)}{I(z)} = \frac{\lambda G(z)}{1+\lambda G(z)} \tag{5.14}$$

The poles of the system, determining the stability are equal to the roots of the stability equation:

$$1+\lambda G(z) = 0 \tag{5.15}$$

The roots of this equation depend on the value of λ, and the stability of the SDM is determined by their position in the complex z-plain:

1. All roots are inside the unit-circle: The SDM is stable.
2. One or more roots lie outside the unit-circle: The SDM is unstable.
3. One or more roots are on or outside the unit-circle and will move inside in case the signal within the loop increases (and λ decreases): The modulator can exhibit stable limit cycles.
4. One or more roots are on or inside the unit-circle and will move outside in case the signal within the loop increases (and λ decreases): The modulator contains an unstable limit cycle. The modulator is on the verge of becoming unstable or entering a stable large-signal limit cycle.

In the following subsections some examples are used to show that the root locus method, using the model for the quantizer described above, fails to predict idle patterns and large-signal limit cycles.

5.3.1 Second Order Lowpass SDM

First, a second order modulator will be examined. The loop filter of the SDM is described by

$$G(z) = \frac{2z^{-1}-z^{-2}}{1-2z^{-1}+z^{-2}}. \tag{5.16}$$

The NTF of this modulator equals

$$\mathrm{NTF}(z) = \frac{1}{1+G(z)} = (1-z^{-1})^2. \tag{5.17}$$

The NTF reveals the lowpass ($z=1$) error-shaping characteristic. Drawing the position of the roots of (5.15) in the complex plane for $\lambda=0$ to $\lambda=\infty$ results in the root locus

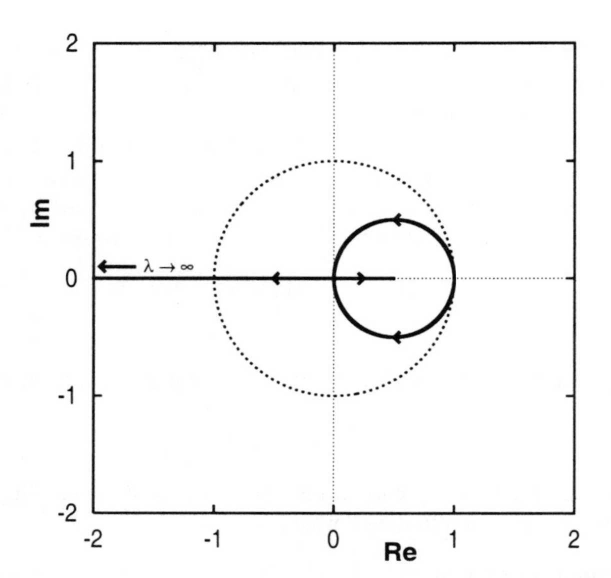

Figure 5.4: *Root locus of a second order lowpass SDM with loop filter (5.16) and the quantizer modeled by a single gain* $\lambda \in [0, \infty)$.

shown in Fig. 5.4 (arrows indicate increasing value for λ). The root locus reveals an idle pattern: at $z = -1$ (half the sample frequency f_s) a root enters the unit circle if λ is decreased. Because λ is relatively large, the signal amplitude at the input of the quantizer will be small and the limit cycle may be considered an idle pattern.

However, simulations and measurements using the experimental set-up described in chapter 7 reveal that such a second-order system generates a "0011" idle pattern under zero input condition, corresponding to a frequency of $f_s/4$. Even at small input levels the modulator prefers coding of the signal with a quarter of the sample frequency, a behavior not predicted accurately by the model used (see Fig. 4.9).

5.3.2 Third Order Lowpass SDM

In this section, the root locus of a third-order lowpass SDM will be analyzed. The loop filter of the modulator is given by:

$$G(z) = \frac{3z^{-1} - 3z^{-2} + z^{-3}}{(1 - 0.95 \cdot z^{-1})^3}. \tag{5.18}$$

The poles of the loop filter are placed near but not on the unit circle for the purpose of the explanation. The underlying reason for this choice will be explained below. The loop filter transfer is given in discrete time domain by:

$$\begin{aligned} y[k] &= -2.85 \cdot y[k-1] + 2.7075 \cdot y[k-2] - 0.8574 \cdot y[k-3] + \\ &\qquad 3 \cdot x[k-1] - 3 \cdot x[k-2] + x[k-3] \end{aligned} \tag{5.19}$$

in which $y[k]$ is the output and $x[k]$ is the input of the loop filter. The root locus of the third order SDM with loop filter (5.18) is shown in Fig. 5.5. Apart from the idle

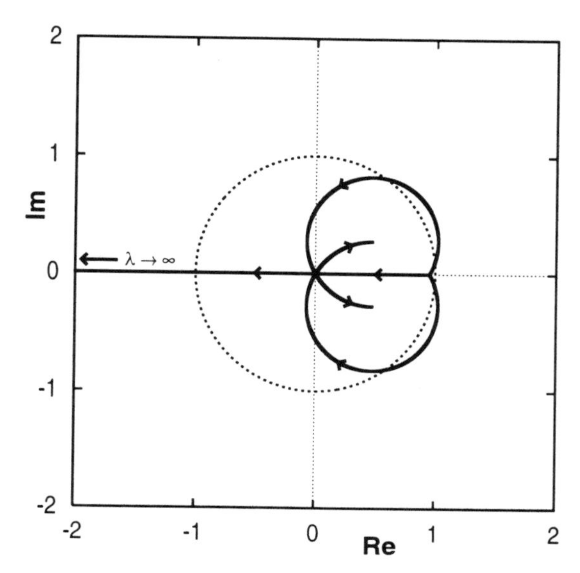

Figure 5.5: Root locus of a third order lowpass SDM with loop filter (5.18) and the quantizer modeled by a single gain $\lambda \in [0, \infty\rangle$.

pattern at $z = -1$ the root locus reveals an unstable limit cycle near $z = e^{\pm j\frac{\pi}{3}}$ and a stable large-signal limit cycle near $z = 1$. For the unstable limit cycle, the roots leave the unit circle if the gain λ decreases below a certain value λ_0. Once the roots are located outside the unit circle, the amplitude of the signal inside the loop will continue to increase, and the behavior of the modulator will converge to the stable large-signal limit cycle near $z = 1$. The amplitude (at the input of the quantizer) of the large-signal limit cycle can be calculated as follows. The output of the quantizer is assumed to be a square wave with amplitude q. The amplitude of the fundamental harmonic can be determined by expanding the square wave into a Fourier series, which gives $4q/\pi$. As λ represents the gain of the fundamental harmonic by the quantizer, the amplitude of the signal at the input of the quantizer is given by $A_{\text{in}} = 4q/(\pi\lambda)$. Here, the predicted amplitude A_{lc} and angular frequency θ_{lc} of the large-signal limit cycle equals:

$$A_{\text{lc}} \approx \frac{\frac{4}{\pi} \cdot \frac{q}{2}}{\lambda} = 0.971 \cdot 10^3 \qquad \text{and:} \qquad \theta_{\text{lc}} \approx 0.1 \text{ rad} \tag{5.20}$$

This large-signal limit cycle is not expected to appear under zero input conditions. Starting from zero initial state ($\lambda = \infty$), a signal with increasing amplitude will appear inside the loop, and λ will decrease. For a certain value $\lambda_1 > \lambda_0$ all roots will be inside or on the unit circle, and an idle pattern (at $z = -1$) with stable amplitude is to be expected. Simulations and experiments show that even under zero input conditions the large signal limit cycle near $z = 1$ is entered and the states of the system become very large. In Fig. 5.6 the output $x[n]$ of the loop filter is shown under zero input condition for a SDM with a loop filter described by (5.18). The loop filter output $x[n]$ reveals a large signal limit cycle. The amplitude of the large signal limit cycle is

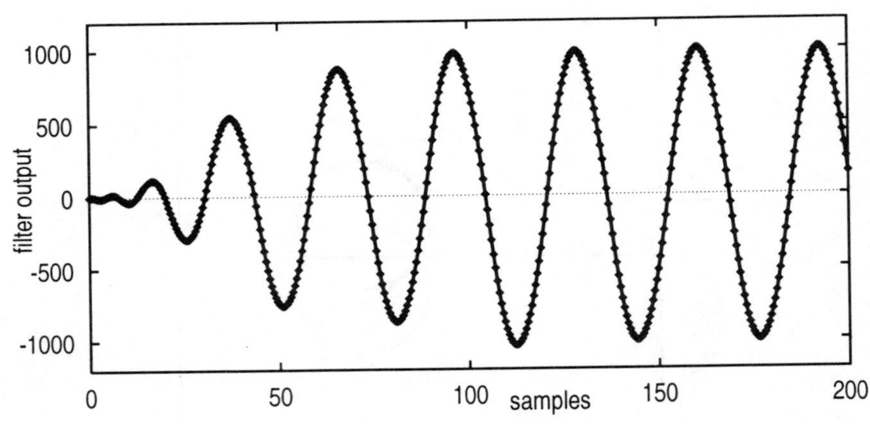

Figure 5.6: Loop filter output of a third order lowpass SDM with loop filter (5.18), zero input and zero initial states. Quantizer step size $q = 2$.

approximately $A_{lc} \approx 1002$ (see Fig. 5.6). The period of the large signal limit cycle is equal to 64 sample periods, corresponding to an angular frequency of $\theta_{lc} = \pi/32 \approx 0.098$ rad. The predictions for the angular frequency and amplitude of the large-signal limit cycle obtained from the root locus are reasonable estimates of actual values. However, the df method did not predict the occurrence of this limit cycle for zero initial state conditions. In the case that the poles of the loop filter were placed exactly on the unit circle, the value for λ for the large-signal limit cycle at $z = 1$ would be equal to $\lambda = 0$. The amplitude of the large signal limit cycle would be infinite. In practice, the states of the third order modulator with the poles placed in the unit circle, grow continuously and do not appear to be bounded. In strict sense, this behavior is not a large-signal limit cycle.

5.3.3 Quantizer Modeling

As was shown in the previous section, modeling of the quantizer by a single global gain is not adequate for prediction of certain aspects of the behavior of the SDM. The idle pattern of a second order lowpass modulator and the occurrence of a large-signal limit cycle in a third order modulator under zero initial state conditions is not predicted correctly by the df method. In the following section it will be shown that a sampled quantizer also exhibits a phase shift that has a significant impact on the stability properties of sigma delta modulators.

5.4 Phase Uncertainty of a Sampled Quantizer [54]

Stability of linear systems with a feedback loop is governed by the amplitude and phase transfer of the elements of the loop. For the df method a nonlinear element is modeled by a (signal dependent) linear transfer. For correct analysis of the stability of the nonlinear feedback loop, a possible phase shift of the fundamental harmonic resulting from the nonlinear element should also be modeled.

In this section the phase shift of a sampled quantizer will be investigated. The phase shift introduced by the quantizer can be considered a phase uncertainty. Sampling of the input signal of the quantizer causes a quantization threshold crossing to be detected by the sample moment *following* this crossing. As the crossing could have occurred anywhere in the previous sampling period, an uncertainty in the phase of the signal is introduced. This phase uncertainty, which depends on the frequency of the input signal, was first mentioned by HÖFELT [49], who used it to verify the existence of limit cycles in SDMs using TSYPKIN's method.

As an example, Fig. 5.7 shows the phase uncertainty of an input sine wave with frequency $f_s/4$ in the case of a one-bit quantizer. The input signal is depicted by a solid line and the output samples are represented by impulses. Clearly, the input signal can be shifted in phase without changing the sign of the signal at the sample moments. Even a considerable phase shift of $\pm\frac{\pi}{4}$ does not affect the output samples.

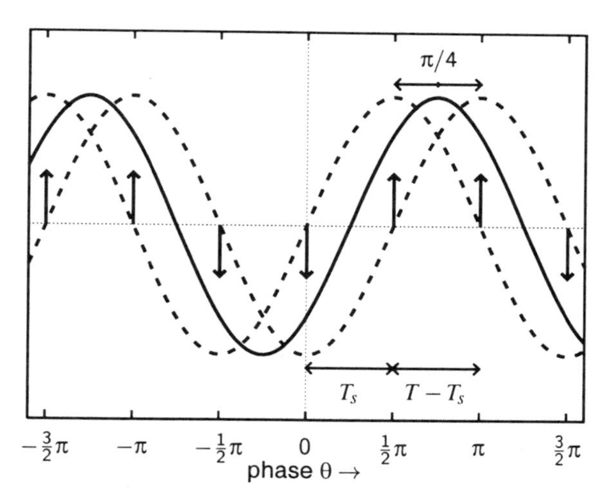

Figure 5.7: *Phase uncertainty of a one bit quantized and sampled sine wave with frequency* $f_s/4$.

In the following sections, the phase uncertainty of sampled quantizers will be examined more closely. Closed-form expressions for the phase uncertainty of a one bit and a two bit quantizer will be derived. In order to construct an acceptable model for the df method, an approximation scheme for the phase uncertainty of a one bit quantizer will also be derived.

5.4.1 Analysis

A sampled quantizer has quantization output levels $q_{o,p}$ and quantization threshold levels q_p. In the case of uniform quantization with quantization step size q, the quantization threshold levels are equal to

$$q_p = q \cdot p \qquad \text{with:} \quad q > 0 \qquad \text{and} \quad p \in \{-P, \ldots, P\} \tag{5.21}$$

in which $2P+1$ is the total number of quantization threshold levels. Consequently, the total number of quantization output levels is $2P+2$. The quantization output levels $q_{o,p}$ are also considered to be uniformly distributed with step size q. The quantizer is sampled at the sample moments t_k. In the case of uniform sampling, the sample moments are equal to

$$t_k = k \cdot T_s \qquad \text{with:} \quad k \in \mathbb{Z} \tag{5.22}$$

in which $T_s = 1/f_s$ is the sampling period. Finally, let the (continuous time) input signal of the quantizer be a sine wave

$$x(t) = A \cdot \sin(2\pi f t + \phi) \tag{5.23}$$

with an amplitude of $A > 0$, a frequency of f and a phase of ϕ.

The phase uncertainty of the sampled quantizer can now be determined as follows. A quantization threshold crossing by the input signal is detected without any phase error when the input sine wave is sampled exactly at this crossing. In other words, the input sine wave is equal to a quantization threshold level q_p at any of the sample moments t_k:

$$A \cdot \sin(2\pi f t_k + \phi) = q_p \tag{5.24}$$

In the case of uniform quantization and uniform sampling this changes to:

$$A \cdot \sin(2\pi f k T_s + \phi) = p \cdot q \tag{5.25}$$

Solving the phase ϕ from (5.25) results in a set Φ of phases of the input signal for which the phase uncertainty is zero. Note that equation (5.25) has solutions for ϕ for $p \in \{-P^\star, \ldots, P^\star\}$ with $P^\star = \lfloor \frac{A}{q} \rfloor$ equal to the index of the highest quantization threshold level reached by the input signal. Without loss of generality, the input signal can be assumed to cross all quantization threshold levels, i.e. $P^\star = P$. For the set Φ it holds that

$$\Phi: \quad \phi = \arcsin(\frac{pq}{A}) + 2\pi k \frac{f}{f_s} + l\pi \quad \text{with:} \left\{ \begin{array}{l} k, l \in \mathbb{Z} \\ p \in \{-P^\star, \ldots, P^\star\} \end{array} \right. \tag{5.26}$$

Any phase $\phi \notin \Phi$ of the input signal will result in a phase error $\phi - \phi^\star$, with $\phi^\star \in \Phi$ which is nearest to ϕ. As a result, the maximum absolute phase error or phase uncertainty equals half the maximum distance between two adjacent solutions $\phi_1, \phi_2 \in \Phi$:

$$\Delta\phi_{\max} = \frac{1}{2} \max(\phi_2 - \phi_1) \quad \text{with:} \left\{ \begin{array}{l} \phi_2 > \phi_1 \\ \langle \phi_1, \phi_2 \rangle \cap \Phi = \emptyset \end{array} \right. \tag{5.27}$$

Equation (5.26) shows that the maximum phase uncertainty depends on the ratio of the input frequency and the sample frequency f/f_s. As the solutions are discrete with respect to the ratio f/f_s, a distinction will be made between input frequencies that are equal to a rational fraction of the sample frequency, and those that are not.

A) Input frequencies not equal to a rational fraction of the sample frequency

The phase uncertainty of the sampled quantizer is zero for input frequencies that are not a rational fraction of the sample frequency, i.e. $f/f_s \in \mathbb{R} \setminus \mathbb{Q}$. This can be shown by reviewing eq. (5.26). According to [55], any set $n + m\xi$ with m, n arbitrary integers and ξ an irrational number will be dense in $\mathbb{R}$. As we assumed f/f_s to be an irrational number, set Φ of solutions of (5.26) will also be dense in $\mathbb{R}$. As a result, the difference between any two adjacent solutions and the maximum phase uncertainty $\Delta\phi_{\max}$ is zero.

In the case that the input frequency is not a rational fraction of the sample frequency, any change of the phase of the input sine wave results in a change of at least one of the output samples. In other words, there will always be a sample moment in time at which the input signal will be equal to a quantization-threshold level. As a result, any change of the phase will be "detected" by the sampled quantizer.

B) Input frequencies equal to a rational fraction of the sample frequency

If the input frequency f is a rational fraction of the sample frequency f_s, this fraction can be written as

$$\frac{f}{f_s} = \frac{m}{M} \tag{5.28}$$

with $m, M \in \mathbb{N}$ and $\gcd(m, M) = 1$.[3] In the case that the Nyquist criterion ($f/f_s \leq \frac{1}{2}$) is taken into account, m and M also satisfy

$$\begin{cases} M \geq 2 \\ m \leq \lfloor \frac{M}{2} \rfloor \end{cases} . \tag{5.29}$$

Substitution of (5.28) into (5.26) and using *Bezout's Theorem* [56]:

$$\forall_{m,M \in \mathbb{Z}} \mid \gcd(m, M) = 1 \ \exists_{\alpha_1, \alpha_2 \in \mathbb{Z}} \ \alpha_1 m + \alpha_2 M = 1$$

equation (5.26) can be simplified to

$$\Phi: \quad \phi = \begin{cases} \arcsin(\frac{pq}{A}) + \frac{2\pi}{M} k' & M \text{ even} \\ \arcsin(\frac{pq}{A}) + \frac{\pi}{M} k' & M \text{ odd} \end{cases} . \tag{5.30}$$

The phase uncertainty for input frequencies satisfying (5.28) approaches zero for an infinite number of quantization-level crossings. Set Φ in (5.30) can be regarded as a repeatedly transposed subset Φ_0. The transposition equals $2\pi/M$ (M even) or π/M (M odd). Φ_0 is defined by

$$\Phi_0: \quad \phi = \arcsin(\frac{pq}{A}) \quad p \in \{-P^\star, \ldots, P^\star\}. \tag{5.31}$$

In order to find an upper bound for the phase uncertainty, the largest transposition of π is considered. In order to show that the phase uncertainty $\Delta\phi_{\max} \to 0$ for $P \to \infty$, the envelope of a set needs to be defined:

[3] $\gcd(m, M)$ represents the *greatest common divisor* of m and M.

The *envelope* ε of a set Φ is defined as the smallest closed interval encompassing all $\phi \in \Phi$, and is given by

$$\varepsilon\{\Phi\} = [\min_{\phi}(\Phi), \max_{\phi}(\Phi)]. \tag{5.32}$$

As the envelope of the subset Φ_0 is encompassed by $[-\pi/2, \pi/2]$ and the transposition of the subsets equals π, the envelopes of all transposed subsets are disjunct. As a result, the largest interval between two adjacent $\phi \in \Phi$ occurs between the highest element in Φ_0 and the lowest element in $\Phi_0 + \pi$, for the largest sine wave amplitude A resulting in P quantization-level crossings:

$$\max(A)|_{\lfloor \frac{A}{q} \rfloor = P} = [q(P+1)]^{-}. \tag{5.33}$$

Here $[x]^{-}$ represents the largest $y \in \mathbb{R}$ for which $y < x$. As a result, an upper bound for the maximum phase uncertainty is given by

$$\Delta\phi_{\max} \leq \frac{\pi}{2} - \arcsin\left(\tfrac{P}{P+1}\right). \tag{5.34}$$

If the number of quantization-level crossings goes to infinity ($P \to \infty$), the upper bound approaches zero. Because the actual maximum phase uncertainty lies between zero and this upper bound, the phase uncertainty also decreases to zero. This follows with

$$0 \leq \lim_{\lfloor \frac{A}{q} \rfloor \to \infty} \Delta\phi_{\max} \leq \lim_{P \to \infty} \tfrac{\pi}{2} - \arcsin\left(\tfrac{P}{P+1}\right) = 0. \tag{5.35}$$

Note that this does not imply that the maximum phase uncertainty decreases with every increase of the number of quantization levels.

5.4.2 Closed Form Expressions

For the purpose of a model of the quantizer for the stability analysis, closed form expressions should be derived for the maximum phase uncertainty $\Delta\phi_{\max}$. Here closed form expressions will be derived for input frequencies that are a rational fraction of the sample frequency and satisfy (5.28) and (5.29) in the case of a one bit and a two bit quantizer.

A) One bit quantizer

A one bit quantizer has one quantization threshold level at $p = 0$. As a result, equation (5.30) is reduced to

$$\Phi: \quad \phi = \begin{cases} \frac{2\pi}{M}k & M \text{ even} \\ \frac{\pi}{M}k & M \text{ odd} \end{cases} \quad k \in \mathbb{Z}. \tag{5.36}$$

The maximum phase uncertainty can be derived directly from (5.27). This gives

$$\Delta\phi_{\max}(f) = \begin{cases} \frac{\pi}{M} & M \text{ even} \\ \frac{\pi}{2M} & M \text{ odd} \end{cases} \quad \text{with:} \quad \frac{f}{f_s} = \frac{m}{M} \quad \text{and} \quad m, M \in \mathbb{N} \tag{5.37}$$

From (5.37) it follows that the maximum phase uncertainty does not depend on the sine wave amplitude, as could be expected from the presence of a single quantization threshold level. In Fig. 5.8 the phase uncertainty is shown for $M = 3, \ldots, 64$.

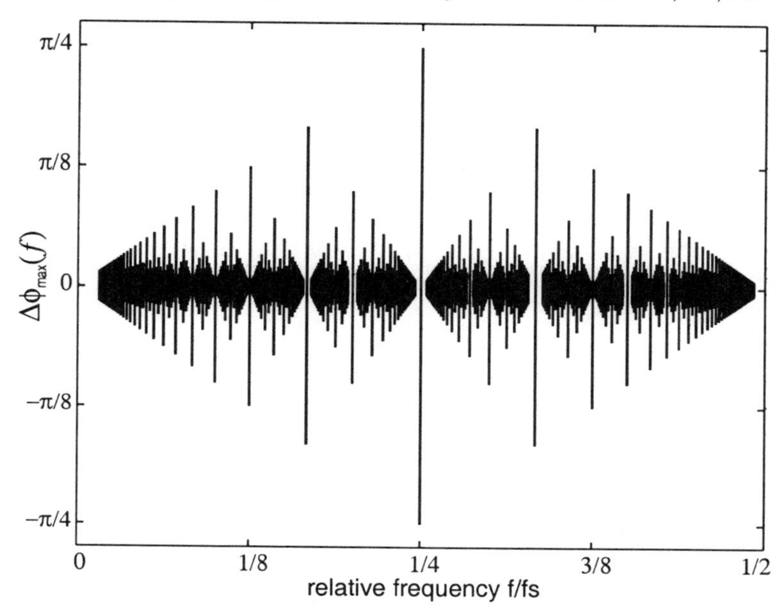

Figure 5.8: *Phase uncertainty of a one bit quantizer.*

In the case that the input frequency equals half the sample frequency ($M = 2$), the maximum phase uncertainty for a single bit quantizer equals $\pi/2$. It should be noted that for a *sampled* sine wave with frequency $f = f_s/2$ a phase shift is indistinguishable from a change in amplitude, regardless of quantization. As a result, the phase uncertainty is undetectable in the case that only the output samples of the quantizer are considered.

B) Two bit quantizer

A two bit quantizer has three quantization levels at $p \in \{-1, 0, 1\}$. Defining ϕ_A as the phase difference between a zero crossing and an adjacent $p = 1$ level crossing

$$\phi_A = \arcsin(\tfrac{q}{A}) \tag{5.38}$$

the solutions for zero phase uncertainty can be written as a repeatedly transposed subset $\{-\phi_A, 0, \phi_A\}$:

$$\Phi: \quad \phi = \{-\phi_A, 0, \phi_A\} + \begin{cases} \frac{2\pi}{M}k & M \text{ even} \\ \frac{\pi}{M}k & M \text{ odd} \end{cases} \tag{5.39}$$

with $k \in \mathbb{Z}$. Because the actual adjacent solutions depend on the values ϕ_A and M, the maximum phase uncertainty has to be determined in four cases. For M even the cases are:

1. The envelopes of repeatedly transposed subsets are disjunct. The distribution of the solutions in (5.39) is depicted in Fig. 5.9a. Each separate transposition ($k = 0, 1, 2, \dots$) of the set $\{-\phi_A, 0, \phi_A\}$ is marked by a different symbol ($+, \Diamond, \Box$ or $\times$). The envelopes are disjunct for $2\pi/M > 2\phi_A$. As a result, the set Φ of solutions satisfies

$$\Phi : \{..., -\tfrac{2\pi}{M} + \phi_A, -\phi_A, 0, \phi_A, \tfrac{2\pi}{M} - \phi_A, \tfrac{2\pi}{M}, \tfrac{2\pi}{M} + \phi_A, ...\}. \tag{5.40}$$

The resulting maximum phase uncertainty, equal to half the maximum distance between consecutive solutions, equals

$$\Delta\phi_{\max} = \begin{cases} \tfrac{1}{2}\phi_A & 2\phi_A < \tfrac{2\pi}{M} \leq 3\phi_A \\ \tfrac{\pi}{M} - \phi_A & 3\phi_A < \tfrac{2\pi}{M} \end{cases} \tag{5.41}$$

2. The envelopes of two adjacent subsets intersect ($\phi_A < 2\pi/M \leq 2\phi_A$). The distribution of solution subsets is shown in Fig. 5.9b. The set Φ of solutions written in ascending order equals

$$\Phi : \{..., -\phi_A, -\tfrac{2\pi}{M} + \phi_A, 0, \tfrac{2\pi}{M} - \phi_A, \phi_A, \tfrac{2\pi}{M}, \tfrac{2\pi}{M} + \phi_A, ...\}. \tag{5.42}$$

Here the maximum phase uncertainty equals

$$\Delta\phi_{\max} = \begin{cases} \tfrac{\pi}{M} - \tfrac{1}{2}\phi_A & \tfrac{3}{2}\phi_A < \tfrac{2\pi}{M} \leq 2\phi_A \\ \phi_A - \tfrac{\pi}{M} & \phi_A < \tfrac{2\pi}{M} \leq \tfrac{3}{2}\phi_A \end{cases} . \tag{5.43}$$

3. The envelopes of three subsets intersect ($\phi_A/2 < 2\pi/M \leq \phi_A$). Figure 5.9c shows the solutions of three subsets. The set Φ of solutions satisfies

$$\Phi : \{..., -\phi_A, -\tfrac{2\pi}{M}, \tfrac{2\pi}{M} - \phi_A, 0, -\tfrac{2\pi}{M} + \phi_A, \tfrac{2\pi}{M}, \phi_A, ...\}. \tag{5.44}$$

The maximum phase uncertainty is described by

$$\Delta\phi_{\max} = \begin{cases} \tfrac{2\pi}{M} - \tfrac{1}{2}\phi_A & \tfrac{2}{3}\phi_A < \tfrac{2\pi}{M} \leq \phi_A \\ \tfrac{1}{2}\phi_A - \tfrac{\pi}{M} & \tfrac{1}{2}\phi_A < \tfrac{2\pi}{M} \leq \tfrac{2}{3}\phi_A \end{cases} . \tag{5.45}$$

4. Four or more envelopes of subsets intersect ($0 < 2\pi/M \leq \phi_A/2$). Here the size of the transposition is smaller than half the interval between two solutions in the same subset, and any interval will be smaller or equal to $2\pi/M$. Figure 5.9d shows the distribution of the solutions of four subsets. The maximum phase uncertainty is equal to half the transposition step size:

$$\Delta\phi_{\max} = \tfrac{\pi}{M} \quad \tfrac{2\pi}{M} \leq \tfrac{1}{2}\phi_A \tag{5.46}$$

Similar expressions are found for odd values of M. In this case $2\pi/M$ in (5.41) through (5.46) should be substituted by π/M. Equations (5.41) through (5.46), together with the expressions for odd values of M constitute the closed-form expression for the maximum phase uncertainty of a two bit sampled quantizer.

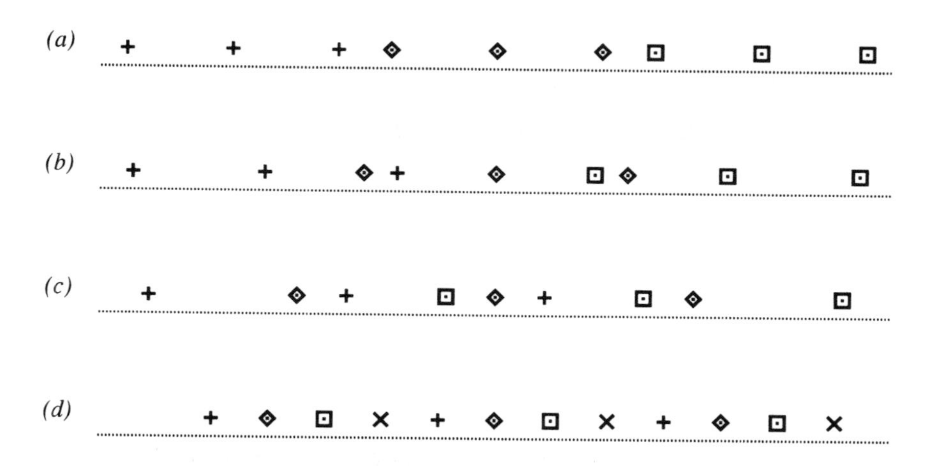

Figure 5.9: Four possible distributions of the zero phase error solution set Φ of a two bit quantizer.

In Fig. 5.10 the maximum phase uncertainty of a two bit sampled quantizer is shown for a sine wave amplitude $A = 1.2q$ and $A = 1.8q$, together with the worst-case maximum phase uncertainty for each frequency.

As is clear from equation (5.46) and Fig. 5.10c, the worst-case value for the phase uncertainty is equal to that of a single bit quantizer. Despite the fact that the input signal amplitude exceeds the quantization threshold q (at $p = 1$), the phase ϕ of the input signal can be such that at every sample moment the signal level is lower than this threshold. In particular, this situation occurs for amplitudes of the input sine wave just exceeding the additional quantization levels at $p = \pm 1$.

5.4.3 Approximation

The discrete character of the maximum phase uncertainty complicates the implementation in a model suitable for the root locus method, in which a continuous phase response is desirable. Also, the maximum phase uncertainty in equation (5.37) relates to an infinite time interval ($k \in \mathbb{Z}$). The phase uncertainty over a short time interval may be considerably larger. Solving the maximum phase uncertainty of a one bit quantizer described by (5.26) and (5.27) for a limited time interval $k \in \{0, \ldots, K\}$ results in an approximation with linear functions. For example, the maximum phase uncertainty determined over a single sample period corresponding to $K = 1$ can be written as:

$$\Delta\phi_{\max}^{K=1}(f) = \begin{cases} \pi \cdot \frac{f}{f_s} & 0 \leq f/f_s \leq \frac{1}{4} \\ \pi \cdot (\frac{1}{2} - \frac{f}{f_s}) & \frac{1}{4} < f/f_s < \frac{1}{2} \end{cases} \tag{5.47}$$

In Fig. 5.11 the approximations for $K = 1$ to $K = 4$ are shown. Clearly, the approximations correspond to the envelope of the actual (discrete) maximum phase uncertainty

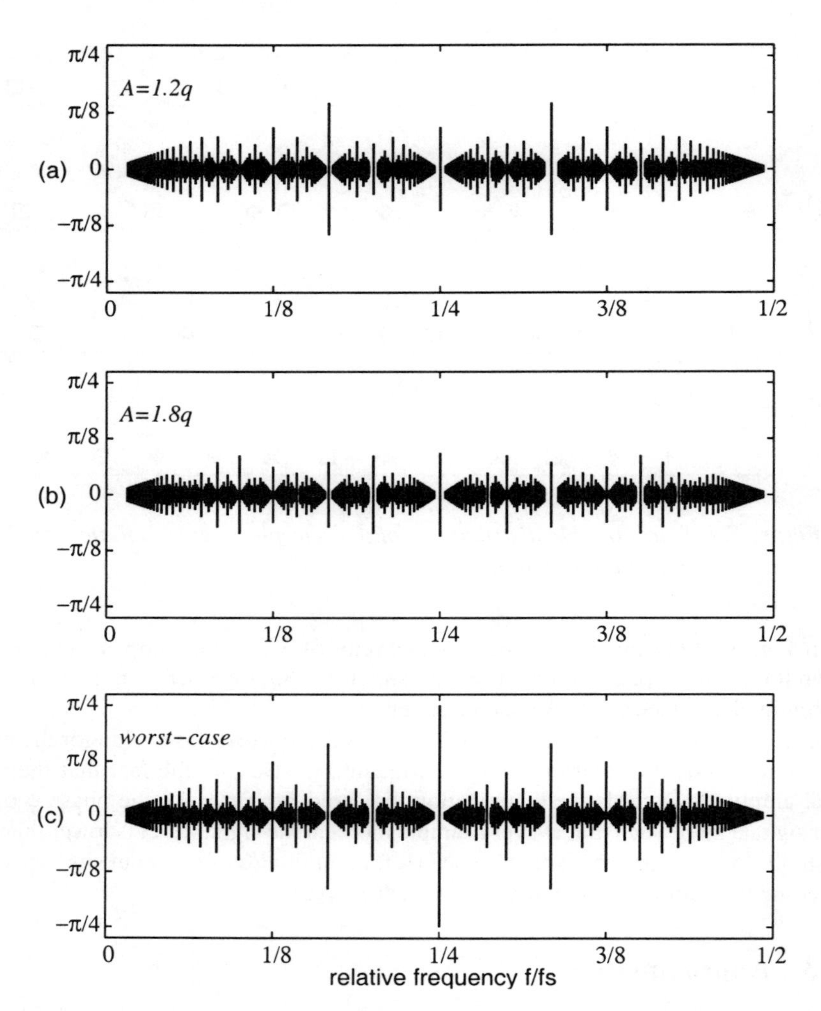

Figure 5.10: Phase uncertainty of a two bit quantizer: (a) for amplitude $A = 1.2q$, (b) for amplitude $A = 1.8q$ and (c) worst-case values.

shown in Fig. 5.8. Using more samples to determine the approximation increases the accuracy, but also complicates the piece wise linear description of the approximation. This approximation scheme thus allows a trade off between complexity and accuracy of the maximum phase uncertainty of a sampled quantizer.

5.4.4 Extended Describing Function Quantizer Model

Using (an approximation of) the maximum phase uncertainty, the linearized model for a sampled (one bit) quantizer can be extended with a phase transfer. The maximum phase uncertainty $\Delta\phi_{max}$ determines the range of the phase uncertainty as a function of

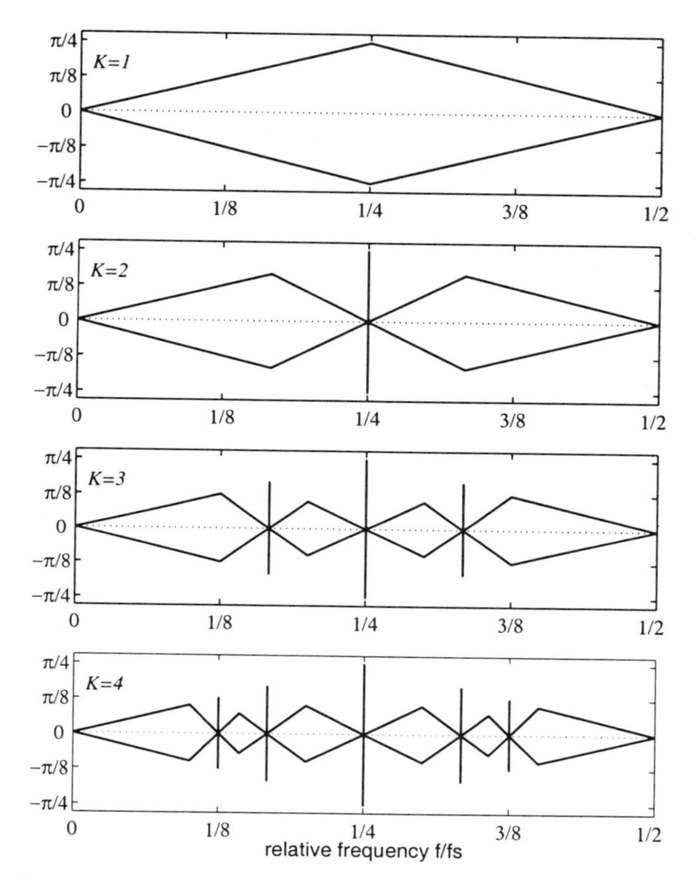

Figure 5.11: *Piecewise linear approximations of the maximum phase uncertainty of one bit quantizer, for several values of K.*

the signal frequency. In order to represent the actual phase uncertainty, a new model parameter α is introduced with $\alpha \in [-1,1]$. The actual phase uncertainty is represented by $\alpha \cdot \Delta\phi_{\max}(\theta)$, in which θ is the normalized angular frequency $\theta = 2\pi f/f_s$. Together with the gain parameter λ which models the effects of amplitude quantization, the linearized z-domain model for a sampled quantizer will be written as:

$$\tilde{Q}(z) = \lambda \cdot e^{j\alpha\Delta\phi_{\max}(\theta)} \quad \text{with:} \quad \begin{cases} z = r \cdot e^{j\theta} \\ \lambda \in [0,\infty\rangle \\ \alpha \in [-1,1] \end{cases} \tag{5.48}$$

Note that for $\alpha = 0$ the model reduces to the original model of $\tilde{Q}(z) = \lambda$ consisting of a gain only. Incorporating a phase transfer in the linear model can therefore be seen as an extension of the linear gain model.

5.5 Prediction of Limit Cycles

The extended describing function model for a sampled quantizer will now be used for the examination of the stability of the sigma delta modulator. In this section the model will be used to predict idle patterns and limit cycles. Prediction of idle patterns is important as they affect the actual dead-zone and in-band tones of the modulator (see sec. 4.2). Figure 5.12 shows the complete linearized model of an SDM to determine the stability. The transfer from the input to the output is given by

$$\frac{O(z)}{I(z)} = \frac{\lambda e^{j\alpha\Delta\phi_{\max}(\theta)} G(z)}{1 + \lambda e^{j\alpha\Delta\phi_{\max}(\theta)} G(z)} \tag{5.49}$$

The poles, determining the stability of this linearized system, are equal to the roots of the stability equation:

$$1 + \lambda e^{j\alpha\Delta\phi_{\max}(\theta)} G(z) = 0 \tag{5.50}$$

5.5.1 Phase Criterion

In order to determine possible limit cycles, the stability equation is reduced to a phase criterion. In the case of a limit cycle, one or more poles of the closed-loop system will be located on the unit circle: $z = e^{j\theta}$. This reduces eq. (5.50) to

$$\lambda e^{j\alpha\Delta\phi_{\max}(\theta)} G(e^{j\theta}) = -1 \tag{5.51}$$

This requirement for the existence of a limit cycle is partitioned into a modulus and a phase requirement:

$$\lambda |G(e^{j\theta})| = 1 \tag{5.52}$$

$$\alpha\Delta\phi_{\max}(\theta) + \mathrm{Arg}\{G(e^{j\theta})\} = \pi \tag{5.53}$$

As the quantizer has an arbitrary positive gain of λ, the modulus requirement (5.52) is always satisfied for some λ. Whether or not a limit cycle with a certain frequency θ is possible, is determined by the phase criterion:

$$\alpha\Delta\phi_{\max}(\theta) + \mathrm{Arg}\{G(e^{j\theta})\} - \pi = 0 \tag{5.54}$$

In a graphical representation, this phase criterion can be verified by drawing the phase shift of the loop filter $\mathrm{Arg}\{G(e^{j\theta})\}$ minus π and applying the phase uncertainty from

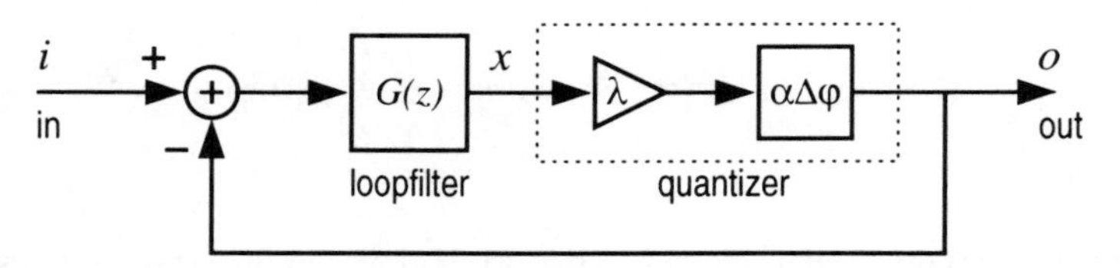

Figure 5.12: *Stability model of a sigma delta modulator.*

Fig. 5.8 as an error-band. In the case that zero phase shift falls within the error-band, the phase criterion is satisfied. First, the phase criterion will be applied to a first order modulator with loop filter:

$$G(z) = \frac{z^{-1}}{1-z^{-1}} \tag{5.55}$$

The plot of the resulting phase criterion is shown in Fig. 5.13. The figure shows that only a frequency of $f_s/2$ satisfies the phase criterion and is therefore the only possible limit cycle. For other frequencies such as $f_s/4$, the phase shift including the phase uncertainty is only marginally larger than zero. A small deviation in the loop filter transfer function causes the phase criterion to be satisfied for these frequencies.

For example, some additional delay in the loop filter may cause a first order modulator to have an idle pattern with a frequency of $f_s/4$ instead of $f_s/2$. An excess delay in the loop filter can be modeled by a pole close to the origin of the complex plane ($z = 0$). The loop filter changes into:

$$G(z) = \frac{z^{-1}}{1-z^{-1}} \cdot \frac{1}{1-d \cdot z^{-1}} = \frac{z^{-1}}{1-(1+d)z^{-1}+dz^{-2}} \tag{5.56}$$

with $0 < d \ll 1$. Figure 5.14 shows the phase criterion plot for $d = 0.2$. Due to the excess delay, the phase criterion is now satisfied for several frequencies. In practical implementations of a first order lowpass modulator, an idle pattern at $f_s/4$ is often observed.

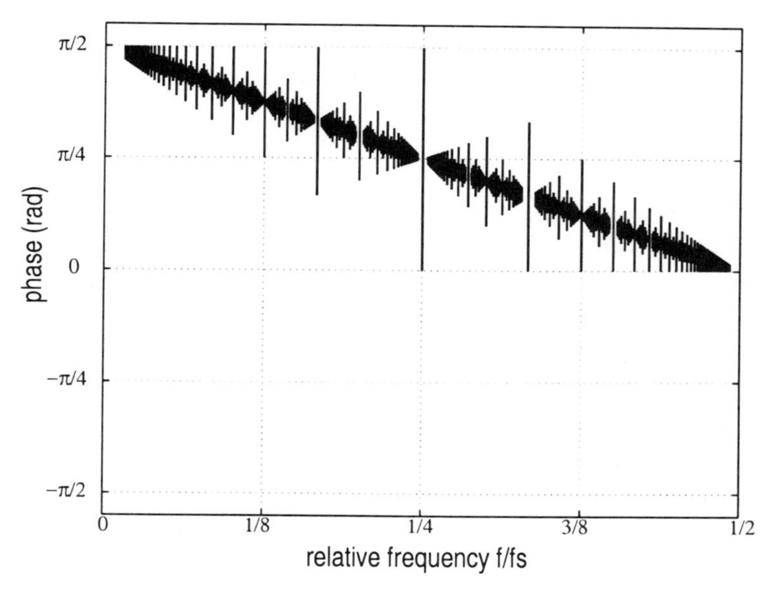

Figure 5.13: *Phase criterion for determination of limit cycles of a first order lowpass modulator.*

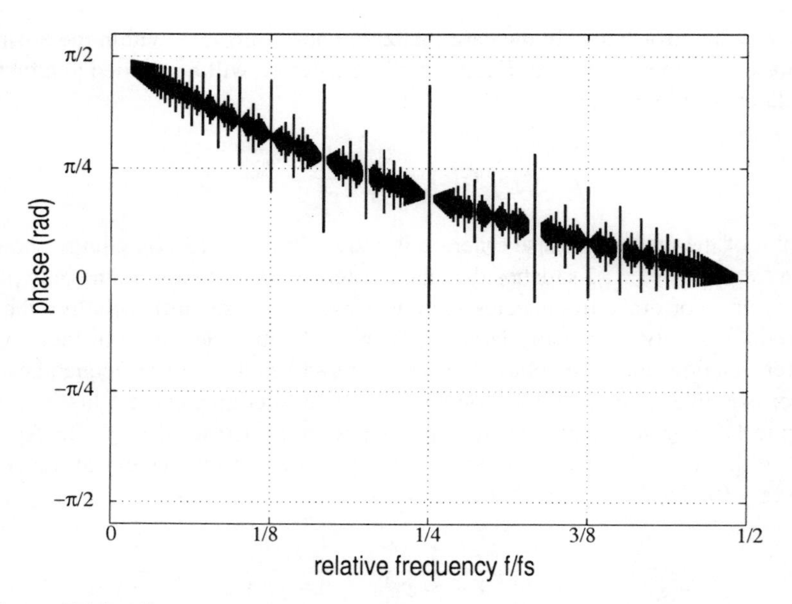

Figure 5.14: *Phase criterion for limit cycles of a first order lowpass modulator with excess delay.*

Application of the phase criterion to the second order lowpass modulator of section 5.3.1 gives Fig. 5.15. From this figure it is clear that several frequencies that are a rational fraction of the sample frequency satisfy the phase criterion and correspond to a possible limit cycle. In addition to $f_{\mathrm{lc}} = f_s/2$ which was already identified as a possible limit cycle, the frequencies that can be written as

$$\frac{f}{f_s} = \frac{\lfloor \frac{M-1}{2} \rfloor}{M} \qquad \text{for:} \quad M = 3,4,5,\ldots \tag{5.57}$$

also satisfy the phase criterion. However, frequencies that correspond to odd M cannot occur in practice. This is caused by the fact that for such frequencies, the output of the quantizer contains a DC value. For example, for $M = 3$ the output of the one-bit quantizer equals a repeated series of either $q/2, q/2, -q/2$ or $-q/2, -q/2, q/2$, giving an average DC value of $\pm q/2$. Generally, limit cycles with frequency f_{lc} will not occur in an SDM with a one-bit quantizer when (aliased) odd multiples of f_{lc} are (nearly) equal to the tuning frequency of the loop filter. The odd harmonic equal to the tuning frequency is amplified significantly by the loop filter and becomes the dominant signal in the loop. As a result, the feedback loop will suppress this harmonic and the fundamental frequency cannot be a limit cycle. Consequently, the limit cycle frequencies of the second order lowpass modulator can be written as

$$f_{\mathrm{lc}} = \frac{1}{2} f_s \quad \text{and:} \quad f_{\mathrm{lc}} = \frac{\frac{M}{2} - 1}{M} f_s \qquad \text{for:} \quad M = 4,6,\ldots \tag{5.58}$$

Note that these frequencies are all located far away from the signal band (at DC) and

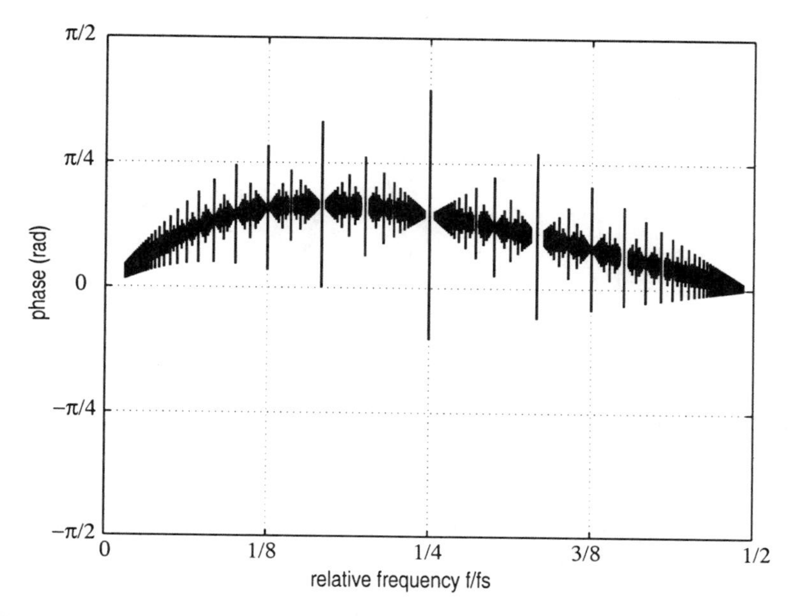

Figure 5.15: *Phase criterion for limit cycles of a second order lowpass modulator.*

the loop filter gain at these frequencies is relatively small. As a result, the limit cycles will have a small amplitude. The second order lowpass modulator can therefore considered to be free of large-signal limit cycles.

Unfortunately, the phase criterion cannot predict which specific limit cycle (or idle pattern) will occur under zero initial state conditions. The criterion can however be used to analyze the behavior of SDMs, and design the modulator (and loop filter in particular) such that it can or cannot exhibit certain limit cycles.

5.5.2 Amplitude and Phase of Limit Cycles

In order to check the prediction of the amplitude and phase of limit cycles, both the modulus eq. (5.52) and the phase eq. (5.53) have to be solved. For the second order lowpass modulator of sec. 5.3.1, the actual idle pattern frequency was $f_s/4$, corresponding to a pair of complex conjugate poles at $z = \pm\, j$. Substitution of either one of these values and solving λ and α gives:

$$\lambda = 0.894 \quad \text{and} \quad \alpha = -0.59 \tag{5.59}$$

Apart from an offset in the output of the loop filter (caused by the initial conditions), these values match the actual values determined from a simulation of the second order modulator (see Fig. 5.16).

The addition of the phase uncertainty to the quantizer model also improves the amplitude prediction of the large-signal limit cycle of the third order lowpass modulator of section 5.3.2. The large-signal limit cycle had a period of $M = 64$, corresponding to

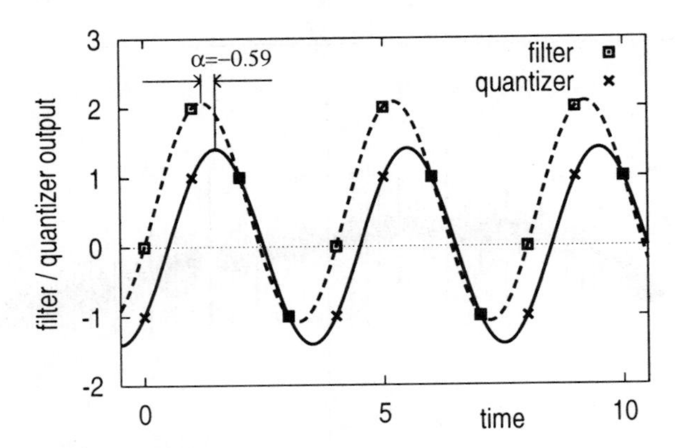

Figure 5.16: *Quantizer and loop filter output showing the idle pattern of a second order lowpass SDM.*

an angular frequency of $\pi/32$. Substituting $z = e^{j\pi/32}$ into the stability equation (5.50) and again solving λ and α gives:

$$\lambda = 1.255 \cdot 10^{-3} \quad \text{and} \quad \alpha = -0.195 \tag{5.60}$$

The resulting (predicted) amplitude of the large-signal limit cycle equals

$$A_{\text{lc}} \approx \frac{\frac{4}{\pi}\frac{q}{2}}{\lambda} = 1.014 \cdot 10^{3} \tag{5.61}$$

which is within 1.5% of the actual value. In order to explain the occurance of the large-signal limit cycle in the case of zero initial state conditions, the root locus of the system needs to be analyzed.

5.6 Small Signal Stability [57]

Caused by the nonlinearity of the sampled quantizer, the stability of an SDM depends on the input signal. An SDM will be called *small signal stable* when its states remain bounded and the modulator is free of large-signal limit cycles for (in-band) input signals with a very small amplitude, i.e. $A_{\text{in}} \ll q/2$. Similarly, an SDM is called *large signal stable* when the modulator is stable for (in-band) input signals with a large amplitude. Note that in both definitions the term "in-band" is used to define the class of input signals. In-band signals are input signals whose frequency (range) is equal to the tuning frequency (range) of the loop filter of the SDM. The response of an SDM to an in-band signal differs considerably from the response to an out-of-band signal, as is shown in an example in Appendix A. As SDMs are mostly used with in-band signals, the stability of the modulator is defined with respect to these signals.

The stability of the SDM will be analyzed by means of the root locus method. Modeling of the phase uncertainty of the sampled quantizer introduces a second parameter to the root locus analysis. As a result, the root locus trajectories that are obtained using a single gain parameter are converted into root locus areas, making the root locus plot hard to evaluate. Therefore, pole trajectories will be plotted as a function of the gain parameter λ, for discrete values of the phase parameter α. The basic thought is that any instability will give rise to a higher amplitude of the signal within the loop and a change in λ. As the phase uncertainty of a one-bit quantizer is independent of the amplitude of the signal, α can considered to be constant. In order to be able to analyze the stability of an SDM with an input signal applied, the following assumption is made: Applying an input signal to the modulator can be modeled by a change in the gain parameter λ. This assumption is supported by the observation that the maximum amplitude at the input of the quantizer indeed depends on the amplitude and frequency of the input signal (see Appendix A).

With respect to the root locus of an SDM, small signal stability can be stated as follows: For any possible value of the phase uncertainty parameter α, there has to be a non-empty range of values for the gain parameter λ for which all the roots of the stability equation of the system reside within the unit circle in the complex z-plane. Starting from $\lambda = \infty$ all root trajectories for constant values of α have to enter the unit circle for decreasing values of λ.

For large-signal stability, the range for which all the poles of the system reside within the unit circle should be large enough to allow variation of λ by the input signal of the modulator. The smallest value of λ for which the poles of the system still reside within the unit circle determines the large-signal stability boundary.

5.6.1 Second Order Lowpass Example

The root locus incorporating the phase uncertainty of the second order lowpass modulator of section 5.3.1 is shown in Fig. 5.17. For the maximum phase uncertainty the approximation determined over a single sample period ($K = 1$) is used as described in section 5.4.3. In that case, the phase uncertainty is given by:

$$\Delta\phi_{\max}^{K=1}(\theta) = \begin{cases} \frac{\theta}{2} & 0 \leq \theta \leq \frac{\pi}{2} \\ \frac{\pi}{4} - \frac{\theta}{2} & \frac{\pi}{2} < \theta \leq \pi \end{cases} \tag{5.62}$$

The root locus can be determined by calculating the roots of:

$$1 + \lambda e^{\alpha \Delta\phi_{\max}^{K=1}(\theta)} G(z) = 0 \quad \text{with:} \quad z = r \cdot e^{j\theta} \tag{5.63}$$

Despite the fact that an approximation is used, the root locus will prove to give a considerable insight in the stability behavior of low- and highpass modulators. Note that equation (5.63) from which the root loci are determined cannot be solved analytically for arbitrary values of λ and α. The root loci shown in this book have been generated using a search method described in Appendix B.

The root locus in Fig. 5.17 of the second order lowpass modulator shows the pole trajectories for several values of the phase uncertainty parameter α. For any value of α, the poles move inside the unit circle when λ decreases. Therefore, this modulator can be considered to be small-signal stable. As the poles remain inside the unit circle for all λ smaller than a certain λ_0 the SDM will also be large signal stable. For $\alpha = -0.59$

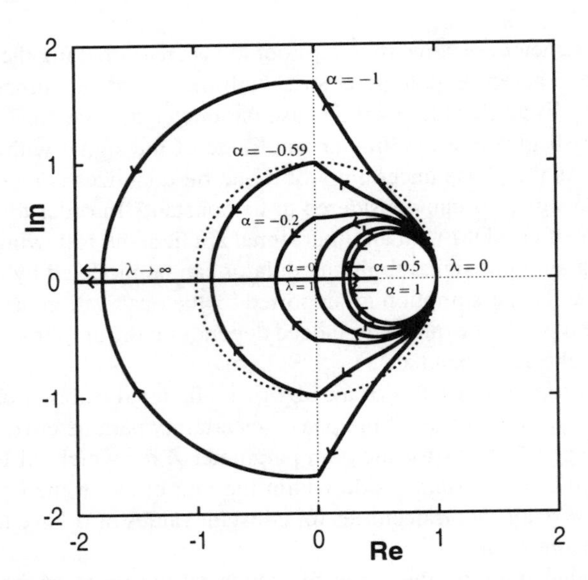

Figure 5.17: Root locus of a second order lowpass SDM described by (5.16) for different values of the phase uncertainty parameter α.

the outer pole trajectory intersects the unit circle at $z = \pm\ j$ corresponding the idle pattern with frequency $f_s/4$. In practice, this idle pattern occurs for zero input conditions and remains dominant at low input signal amplitudes. The idle pattern is slightly disturbed by the input signal. In Fig. 5.18 the output spectrum of the second order modulator is shown for a relative input amplitude of $A_i/q = 3 \cdot 10^{-7}$ and a relative input frequency of $f_i/f_s = 7 \cdot 10^{-5}$.

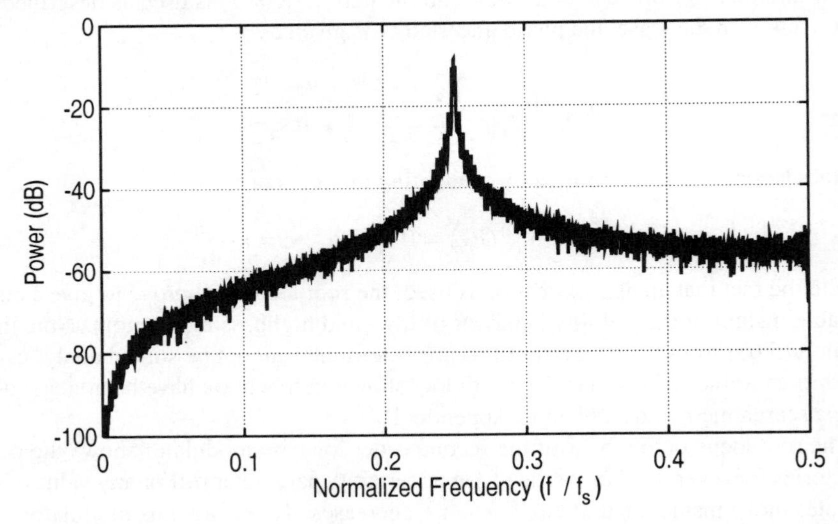

Figure 5.18: Output spectrum of a second order modulator with very low input amplitude (simulation, 2K bins).

5.6.2 Third Order Lowpass Modulator Example

In order to explain why the third order lowpass modulator from section 5.3.2 exhibits a large-signal limit cycle starting from zero initial state conditions, the root locus is determined for several values of the phase uncertainty parameter α. Again the approximation with $K = 1$ is used for the maximum phase uncertainty. Figure 5.19 shows the resulting root locus for $\alpha = \{-1, -0.5, 0.5, 1.0\}$. For $\alpha = 0$ the phase uncertainty is not taken into account and the root locus is identical to the one shown in Fig. 5.5.

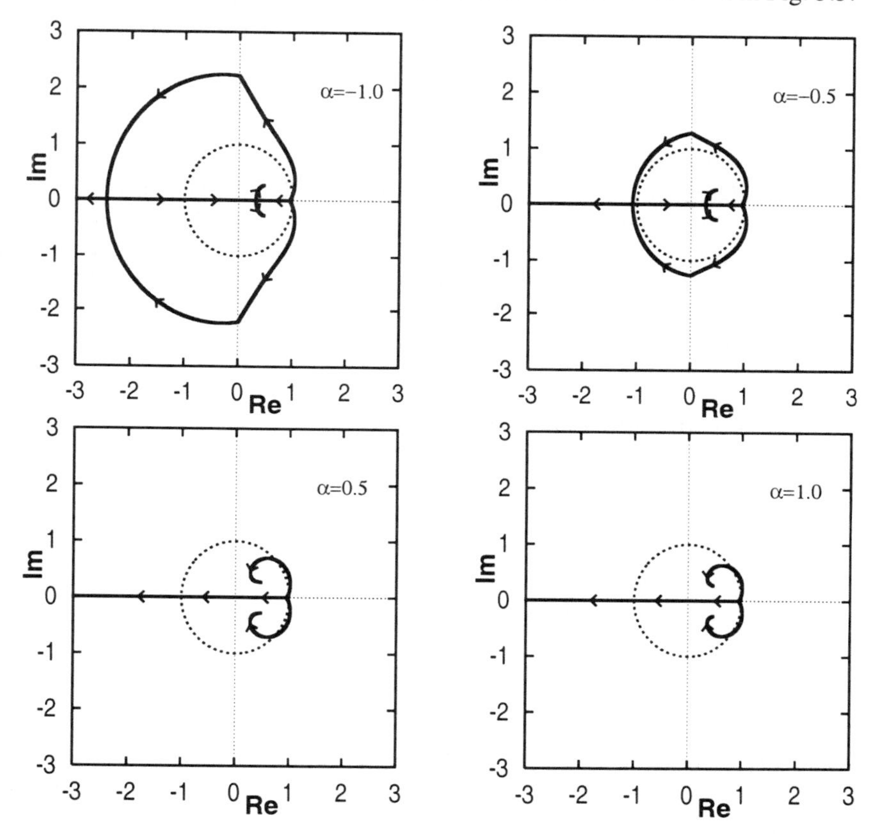

Figure 5.19: Root Locus of third order lowpass SDM described by (5.18) for α=-1,-0.5,0.5 and 1.0.

For near zero initial state conditions, the input of the quantizer will be very small and the gain λ will be very large. For any value of the phase uncertainty parameter α, at least one pole of the closed-loop system will be outside the unit circle when $\lambda \gg 1$. The amplitude of any signal present inside the loop, no matter how small, will increase exponentially. As a result, the signal at the input of the quantizer will become bigger and λ will decrease. For some values of the phase uncertainty parameter α the poles will move inside the unit circle for a moderate value of $\lambda \approx 1$. However, for $\alpha = -1$ (upper left), the poles will not enter the unit circle until λ has become very small ($\lambda \ll 1$) and the poles are near $z \approx 1$. The intersection points with the unit circle represent a stable large-signal limit cycle. In the case that the poles would move inside the unit circle near $z \approx 1$, the signal inside the loop would be damped and it's amplitude would decrease exponentially. Consequently, the gain λ would increase again, moving the poles outside the unit circle. Conversely, when the poles would move outside the unit circle the amplitude of the limit cycle would increase again, lowering λ and pushing the poles back into the unit circle. As a result, the poles settle on the unit circle corresponding to a stable large-signal limit cycle as λ is very small.

This third order lowpass modulator cannot be considered small signal stable, as there is no range of λ for which the SDM is free of large-signal limit cycles *and* all the poles of the loop reside inside the unit circle for any value of the phase uncertainty parameter α.

As was shown in chapter 4, the poles of the loop filter should be placed at the unit circle for optimal noise shaping performance. Even though the poles of the third order SDM considered here lie well within the unit circle and do not give optimal performance, the modulator is not stable. This suggests that small signal stability is not so much determined by the poles of the loop filter, but by the zeroes of the loop filter. In the following section, a class of lowpass modulators will be analyzed and small signal stability boundaries for loop filter parameters will be determined.

5.6.3 Low- and Highpass Modulators

In this section a class[4] of lowpass SDMs will be analyzed having the following loop filter:

$$G(z) = \frac{(1-az^{-1})^N}{(1-bz^{-1})^N} - 1 \quad \text{with:} \quad 0 \leq a,b \leq 1 \tag{5.64}$$

The noise transfer function of these SDMs can be written as:

$$\text{NTF}(z) = \frac{(1-bz^{-1})^N}{(1-az^{-1})^N} \tag{5.65}$$

From this NTF it can be seen that the SDMs have a lowpass noise shaping characteristic for $b \approx 1$. In that case, $|\text{NTF}(z)|$ will be very small for $z \approx 1$ corresponding to very low frequency signals (DC). The parameter b corresponds to the leakage of integrator terms $(1-z^{-1})$. The parameter a determines the location in the complex z-plane of the loop filter zeroes. In order to have good noise shaping performance, a should not be near 1.

By using the transformation $z^{-1} \rightarrow -z^{-1}$, the class of lowpass modulators described by (5.64) can be transformed into a class of highpass modulators. Highpass SDMs will suppress the quantization errors for frequencies near half the sample frequency (corresponding to $z=-1$). As the phase uncertainty in the quantizer model is symmetrical around $f_s/4$, the quantizer model is insensitive to the transformation used to obtain the highpass modulators. Therefore, the analysis of the low- and highpass

[4]Note that the previous examples used in this book also belong to this class.

modulators will give identical results. Here, the analysis will be limited to the class of lowpass modulators.

Figure 5.20 shows the root locus of a third order ($N = 3$) modulator belonging the class of (5.64) with $a = 0$ and $b = 1$ for several values of α. The root locus of the modulator resembles the third order lowpass modulator of the previous section and the SDM cannot be considered small signal stable either. As the poles of this modulator lie exactly on the unit circle at $z = 1$, the modulator does not exhibit a stable large-signal limit cycle. Instead, the amplitude of the signal at the input of the quantizer (corresponding to one of the states of the loop filter) grows continuously.

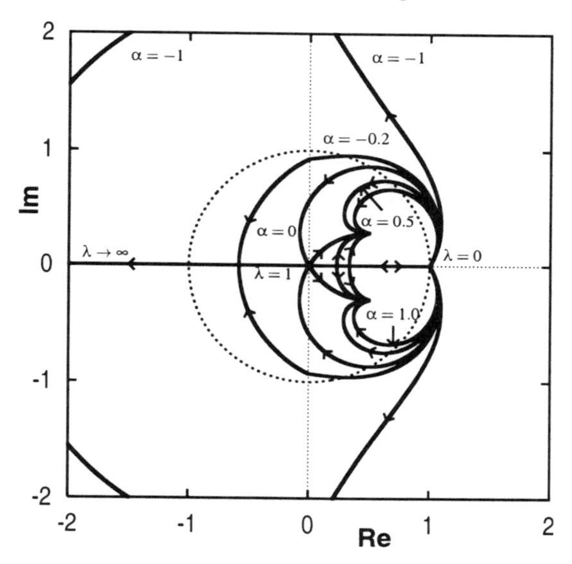

Figure 5.20: Root locus of a third order lowpass SDM ($a = 0$) for α=-1, -0.2, 0, 0.5 and 1.

By changing the loop filter, this third order modulator can be made stable for small input signals. The modulator will be small-signal stable when all the poles of the system reside within the unit circle for some moderate value of λ, regardless the value of α. The value of the loop filter parameter a should be changed such that when λ is decreased from $\lambda = \infty \to 0$, the poles will always enter the unit circle (for some moderate value of λ).

From Figs. 5.19 and 5.20 it follows that the worst-case situation of the phase uncertainty occurs for $\alpha = -1$. This value causes the outermost trajectories to be furthest away from the unit circle. In Fig. 5.21 these outermost branches of the root locus are shown for several values of the filter parameter a. For $a \geq 0.412$ these outermost branches intersect the unit circle between $z = e^{j\pi/3}$ and $z = e^{j\pi/4}$. The poles of the system will enter the unit circle for moderate values of λ for any value of α when λ is decreased: the SDM can be considered stable for small signals. For $a = 0.412$, the outermost branch of the root locus has exactly one intersection point with the unit circle between $z = e^{j\pi/3}$ and $z = e^{j\pi/4}$; the outermost branch is *tangent* to the unit circle.

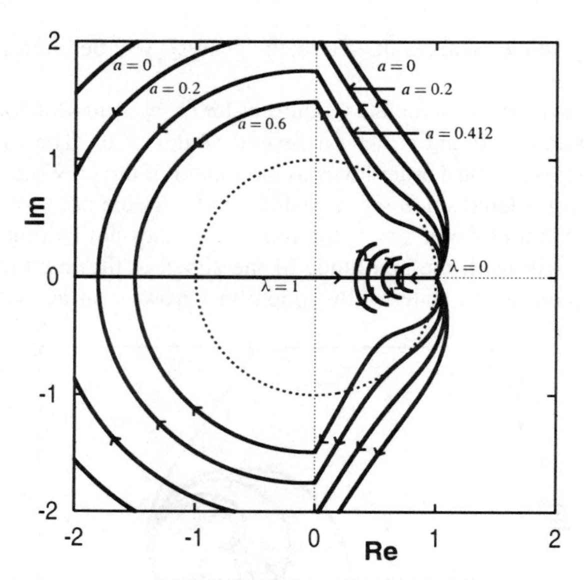

Figure 5.21: *Outermost branches of the root locus ($\alpha = -1$) of a third-order lowpass SDM for several values of the filter parameter a.*

Root loci of lowpass modulators with a higher order ($N \geq 3$) loop filter closely resemble the root locus of the third order modulator shown in Fig. 5.20. Minimum values for a can be obtained in the same way as in the case of the third order SDM. As an example, the root loci of a fourth order and a fifth order modulator are shown in Fig. 5.22 and Fig. 5.23 for $\alpha = 0$ (no phase uncertainty) and $\alpha = -1$ (worst-case phase uncertainty). For the fourth order modulator $N = 4$, $b = 1$ and $a = 0.587$; in the case of the fifth order modulator $N = 5$, $b = 1$ and $a = 0.679$. In these cases, the outermost branches of the root locus are tangent to the unit circle: the SDMs are marginally stable for small signals.

The exact value of the loop filter parameter a for which the root locus intersects with the unit circle can be determined accurately using the algorithm described in Appendix C. The minimum values of a for small signal stability are listed in table 5.1. The small signal stability boundaries were verified using simulations and an all-digital test set-up described in chapter 7. In order to determine the practical stability boundaries the following approach was used. Using an in-band signal with a very small amplitude, the output of the loop filter was examined during the simulation run. If the filter output exceeded a threshold value[5], the SDM was considered to be unstable for small inputs. For the simulations 10^7 samples were calculated using an input signal with frequency $\theta \approx 2 \cdot 10^{-6}$ and amplitude $A \approx 3 \cdot 10^{-4}$. The resulting experimental values are also listed in table 5.1. The theoretical values determined using the df method lie within 5% of the experimental values. These values could be obtained by modeling the phase uncertainty of the quantizer, as the simple linear gain model ($\alpha = 0$) does not suggest instability at low input amplitudes or at zero initial state conditions.

[5] The threshold value was chosen a number of times the quantizer output value (typically 100).

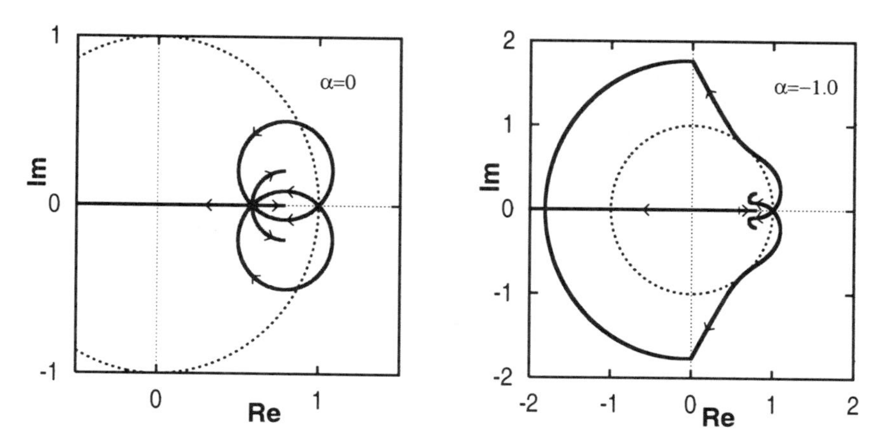

Figure 5.22: Root Locus of fourth order lowpass SDM which is marginally stable for small input signals.

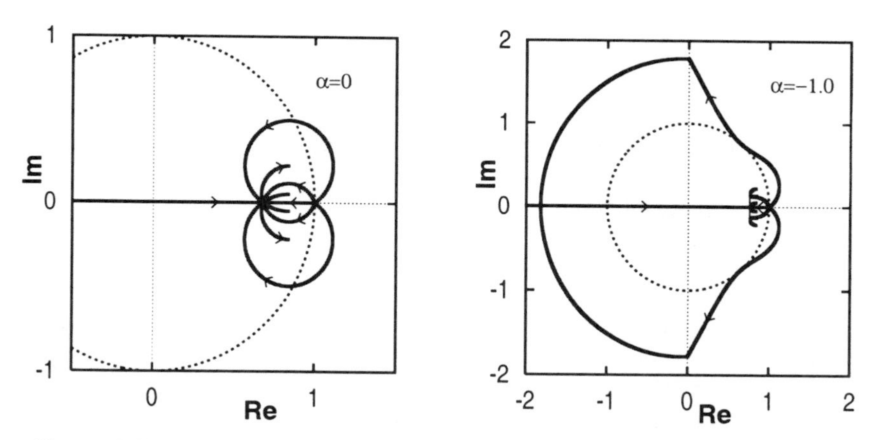

Figure 5.23: Root Locus of fifth order lowpass SDM which is marginally stable for small input signals.

Table 5.1: Minimal values of a for which the system described by (5.64) is stable (see footnote) for small signals for $b = 1$ and order N.

N	a	
	df method	*experimental*[6]
3	0.412	0.416
4	0.587	0.619
5	0.679	0.717
6	0.736	0.771

[6] Here, the SDM was considered to be experimentally stable when it did not become unstable within 10^7 samples (see text).

In order to show that the exact amplitude of the input signal does not affect the experimental results, the true small-signal stability boundary was determined as a function of the input amplitude. Figure 5.24 shows the experimental stability boundary of a third, fourth, fifth and sixth order modulator. For input amplitudes less than -40dB

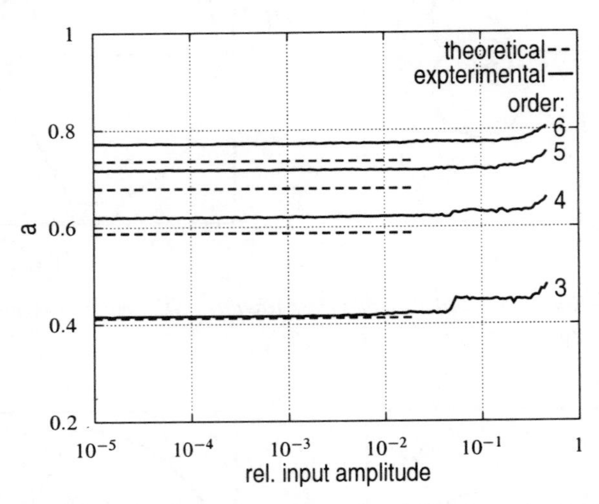

Figure 5.24: *Experimental and theoretical stability boundaries for loop filter parameter a of a SDM with loop filter (5.64) as a function of input amplitude used.*

relative to the quantizer step size, the stability boundary does not change noticeably. For larger input amplitudes, the input signal will affect the input signal of the quantizer and the experimental boundary then agrees with a large-signal stability boundary.

As was mentioned before, the poles of the loop filter do not have a significant effect on the stability of the modulator. In order to verify this observation, the small-signal stability boundaries were determined as a function of the radius of the poles b. In Fig. 5.25 both the values derived using the df method and the experimental values for the small-signal stability boundary of a are shown for a third, fourth and fifth order lowpass SDM. When the poles are moved away from the unit circle ($b < 1$), a less stringent small signal stability boundary can be expected. In other words, the minimum value required for a will decrease when b is decreased. However, the variation of the minimum stability value for a as a function of b is low. This confirms the observation that moving the poles away from the unit circle does not affect the stability behavior of the modulator significantly.

5.6.4 Rule of Thumb

As mentioned in the previous paragraph, the root loci of higher order low- and highpass modulators are similar. This is true even for a modulator not belonging to the class defined by (5.64) that has a loop filter with:

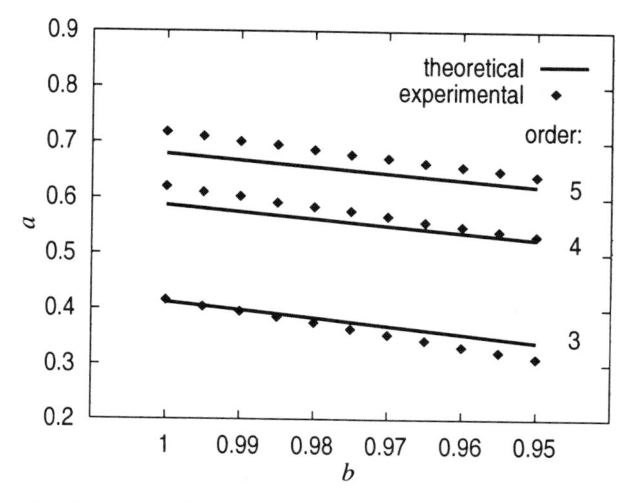

Figure 5.25: Theoretically and experimentally determined minimal values for a of a third- , fourth- and fifth-order lowpass SDM with loop filter (5.64) for which the SDM is stable for small signals.

- the poles near $z = 1$ in the complex plane
- the number of zeroes one less than the number of poles
- the zeroes located in the right half of the unit circle

The root loci of these modulators all have an outermost pole trajectory, the location of which is strongly determined by the phase uncertainty parameter α. The root loci were determined using the approximation with $K = 1$ for the maximum phase uncertainty. This approximation was used in order to have a relatively simple model for the maximum phase uncertainty of the sampled quantizer. The continuous nature of the model results in continuous pole trajectories in the root locus, thereby allowing a better interpretation of the root locus. The actual phase uncertainty has a discrete character (see Fig. 5.8). If a root locus would be plotted using the actual maximum phase uncertainty, interpretation of the root locus would be nearly impossible as the trajectories of the poles would be discontinuous.

The discrete nature of the maximum phase uncertainty can be incorporated into the small-signal stability test when using the following observation. The outermost branch of the root loci of these lowpass modulators are all tangent to the unit circle between $\theta = \pi/4$ and $\theta = \pi/3$, when the approximation with $K = 1$ is used for the maximum phase uncertainty. The discrete maximum phase uncertainty shown in Fig. 5.8 has a local maximum of $\pi/6$ at $\theta = \pi/3$. Instead of requiring the outermost trajectories of the root locus to be tangent to the unit circle, small signal stability can be determined by requiring the outermost poles of the closed-loop system to be inside the unit circle at $\theta = \pi/3$. This leads to the following rule of thumb:

A lowpass modulator with loop filter $G(z)$ will be stable for small input signals when the roots of the stability equation

$$1+\lambda e^{j\alpha\pi/6}G(z)=0 \quad \textbf{with:} \quad z=r\cdot e^{j\theta}$$

lie within the unit circle for $\theta=\pi/3, \lambda>0$ and all $\alpha\in[-1,1]$.

Solving the stability equation for this value of θ gives the radii r of the roots and the corresponding values of λ. By changing parameters of the loop filter $G(z)$, the roots can be moved inside the unit circle, giving a small-signal stability boundary for the loop filter parameters. For highpass modulators, the root loci are mirrored along the imaginary axis in the complex plane. The rule of thumb can be easily adapted for highpass modulators by replacing $\theta=\pi/3$ with $\theta=2\pi/3$.

The rule of thumb will give a slightly more strict stability boundary than the requirement on the pole trajectories being tangent to the unit circle. For example, applying the rule of thumb to a third order SDM with loop filter

$$G(z)=\frac{(z-a)^3}{(z-1)^3}-1 \tag{5.66}$$

results in a 'rule-of-thumb' equation

$$(z-1)^3+\lambda e^{j\alpha\pi/6}\{(z-a)^3-(z-1)^3\}=0 \tag{5.67}$$

As it was found in the previous section, the worst-case phase uncertainty corresponds to $\alpha=-1$. Substitution of $\alpha=-1$ and $z=r\cdot e^{j\pi/3}$ gives

$$(r\cdot e^{j\pi/3}-1)^3+\lambda e^{-j\pi/6}\left\{(r\cdot e^{j\pi/3}-a)^3-(r\cdot e^{j\pi/3}-1)^3\right\}=0 \tag{5.68}$$

from which the radius r (of the roots) is solved. This can be done by setting both the real and imaginary part of the left hand term of (5.68) to zero, as a and λ are real-valued. Although analytic expressions for the solutions of r can be obtained in the case of this third order SDM, such expressions cannot be obtained for higher order modulators. Here, the solutions for r are determined with numerical methods. In general, the solutions for r depend on a. As the rule of thumb requires that the roots are located inside the unit circle, the value for a should be such that $r\leq 1$ for all solutions of (5.68). In the case of the third order SDM this gives: $a\geq 0.449$. Minimum values of a for higher order SDMs can be determined in the same way. The resulting minimum values for a third to sixth order lowpass modulator are listed in table 5.2, together with the values obtained by LEE's rule. Clearly, the values obtained by LEE's rule are more conservative. For the third order SDM in particular, LEE's value deviates considerably from the experimental value, resulting in very conservative design and a corresponding decrease in performance.

5.6.5 Bandpass Modulators

Bandpass modulators have a loop filter which is not tuned at DC or $f_s/2$. The quantization errors of the low resolution quantizer will be suppressed for frequencies close

Table 5.2: Minimal values for a for which the system described by (5.64) is stable for small signals for b = 1 and order N.

N	a			
	rule of thumb	*df method*	*experimental*	*Lee's Rule*
3	0.449	0.412	0.416	0.587
4	0.611	0.587	0.619	0.682
5	0.698	0.679	0.717	0.741
6	0.753	0.736	0.771	0.782

to the tuning frequency θ_0. Often, the tuning frequency of the modulator is chosen equal to a quarter of the sample frequency $f_s/4$, i.e. $\theta_0 = \pi/2$. This is done for several reasons. When the loop filter is implemented using discrete-time circuitry such as switched capacitors or switched current circuits, the loop filter design is simplified as half of the coefficients will be zero and does not have to be implemented. Another reason is that the discrete-time loop filter transfer function can be obtained from a lowpass prototype using the transformation $z^{-1} \rightarrow -z^{-2}$. Similar to the lowpass to highpass transformation, this bandpass transformation preserves the dynamics of the modulator [58]. The stability properties of the resulting bandpass modulators are identical to the properties of the lowpass prototype. This also becomes clear by looking at the root locus of such a bandpass modulator. Transformation of the class of lowpass modulators described by (5.64) gives

$$G(z) = \frac{(1+az^{-2})^N}{(1+bz^{-2})^N} - 1 \tag{5.69}$$

Before calculating a root locus of these modulators, the approximation used for the maximum phase uncertainty should be reconsidered. Comparing the actual maximum phase uncertainty in Fig. 5.8 and the approximation of $K = 1$ in Fig. 5.11 reveals that the approximation of $K = 1$ strongly deviates from the actual phase uncertainty for frequencies near the tuning frequency $f_s/4$. As this could lead to misinterpretation of the root locus, the more accurate approximation $K = 2$ will be used. Figure 5.26 shows the root locus using this approximation of a sixth-order bandpass modulator with a loop filter described by (5.69) with $b = 1$, $a = 0.412$ and $N = 3$. The root locus has been drawn for two values of α. Fig. 5.27 shows a detail of Fig. 5.26(right). The root locus partly resembles a scaled and rotated version of the root locus of the lowpass prototype shown in Figs. 5.20 and 5.21. Although all poles of the closed-loop system will reside within the unit circle for intermediate values of λ, some poles leave the unit circle near $z = e^{j\pi/3}$ and $z = e^{j2\pi/3}$ for moderate values of λ. For $a \geq 0.412$ these outermost pole trajectories of the root locus will re-enter the unit circle before λ becomes really small, thus ensuring small-signal stability of the modulator.

In the root locus of the sixth order bandpass modulator, the distance of the outermost poles to the unit circle has a maximum for $\theta = \pi/3$ and $\theta = 2\pi/3$. By requiring these outermost poles to be within the unit circle for these frequencies, small signal stability can be obtained again. This requirement is identical to a combination of the rule of thumb for lowpass and highpass modulators.

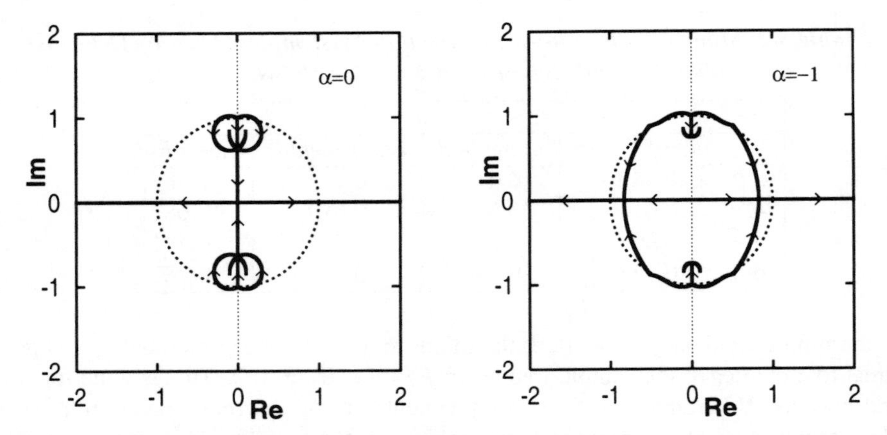

Figure 5.26: *Root Locus of sixth order bandpass SDM which is marginally stable for small input signals.*

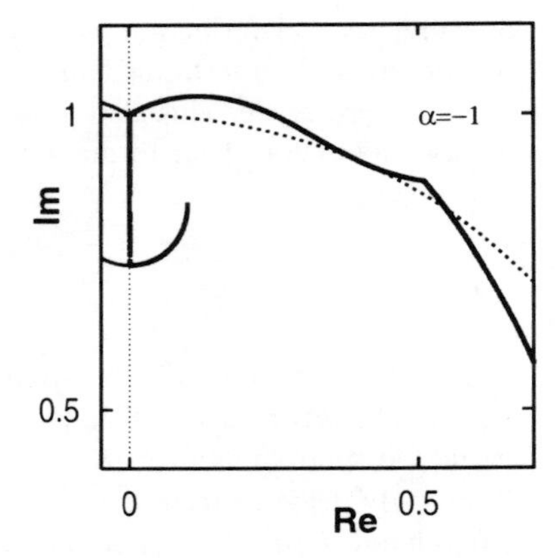

Figure 5.27: *Detail of Fig. 5.26(right).*

Although the rule of thumb was derived for lowpass modulators, it was shown that the rule can be used for highpass modulators and a special class of bandpass modulators after some minor modifications. The SDMs all have similar dynamic behavior and small-signal stability could be determined using a similar rule of thumb. The rule of thumb will now be tested for a more general class of bandpass modulators, having a loop filter given by

$$G(z) = \frac{(1+ae^{j\theta_0}z^{-1})^N(1+ae^{-j\theta_0}z^{-1})^N}{(1+be^{j\theta_0}z^{-1})^N(1+be^{-j\theta_0}z^{-1})^N} - 1 \tag{5.70}$$

For this class of bandpass modulators, θ_0 determines the tuning frequency, b determines the quality factor of the poles and a determines the location of the zeroes. Generally, the poles will be placed near or on the unit circle ($0.95 < b \leq 1$) and the zeroes will be located far away from the poles. For $\theta_0 = 0$, this class of bandpass modulators is equal to the class of lowpass modulators described by (5.64). In the case that $\theta_0 = \pi$, the loop filter is equal to the highpass filters obtained by application of the transformation $z^{-1} \rightarrow -z^{-1}$ to (5.64). For $\theta_0 = \pi/2$, this class of bandpass filters is tuned at $f_s/4$ and equal to the class described by (5.69) in which a has been replaced by a^2 and b has been replaced by b^2. As the bandpass modulators are tunable, both the "lowpass" rule of thumb (i.e. testing at $\theta = \pi/3$) and the "highpass" rule of thumb (testing at $\theta = 2\pi/3$) apply. Figure 5.28 shows the resulting minimum boundaries for a for small-signal stability of a fourth, sixth, eighth and tenth order bandpass modulator with $b = 1$ as a function of the tuning frequency θ_0. The experimentally determined minimum values for a are also shown.

The theoretical boundaries are near the experimental boundaries for $\theta = 0, \pi/2$ and π, as could be expected. For the sixth, eighth and tenth order SDM, the theoretical boundaries follow the experimentally determined boundaries very closely for most of the tuning range. For tuning frequencies within the intervals $[0, \pi/5]$, $[2\pi/5, 3\pi/5]$ and $[4\pi/5, \pi]$ (60% of the tuning range), the rule of thumb provides an accurate estimate of the small-signal stability boundaries (within 5%). Only around $\theta_0 = \pi/3$ and $\theta_0 = 2\pi/3$ a larger deviation occurs. Tuning a bandpass modulator near these frequencies places the poles of the filter on or near the unit circle in the neighborhood of these frequencies. The poles of the closed-loop system (i.e. the roots in the root locus) are near the filter poles for very small values of λ. Consequently, when it is required that the roots of the root locus are located within the unit circle for $\theta = \pi/3$ and $\theta = 2\pi/3$, the stability of the SDM is tested for small values for λ. In that case, the rule of thumb effectively tests the large-signal stability instead of the small signal stability. In other words, for tuning frequencies near $\theta = \pi/3$ and $\theta = 2\pi/3$, the rule of thumb tests the stability of the SDM with a large input signal applied to the modulator. Consequently, the values for a found by the rule of thumb will be higher than the experimental values for small-signal stability (see Fig.5.28).

The fourth order SDM shows considerable deviations between experimental and theoretical boundaries at most of the tuning range. However, the fourth order SDM differs from the higher order modulators in that it becomes large-signal stable for frequencies around $f_s/4$, i.e. $\theta_0 = \pi/2$.

5.6.6 Discussion

In this section a small-signal stability criterion has been derived using the root locus method. The sampled quantizer was replaced by a describing function model that consists of a gain and a phase uncertainty. Analysis of the root loci of a class of lowpass modulators has given insight into the stability behavior of the modulator. Based on this theoretical analysis, a rule of thumb to determine small-signal stability was derived. Other rules of thumb for the stability of SDMs such as LEE's Rule and the Power Gain Rule (see sec. 5.2) are commonly based on empirical results.

Although determined for (a class of) lowpass modulators, the rule of thumb has been applied to a class of tunable bandpass modulators. For modulators of order six and higher, the rule of thumb still provides an accurate estimate for the small-signal

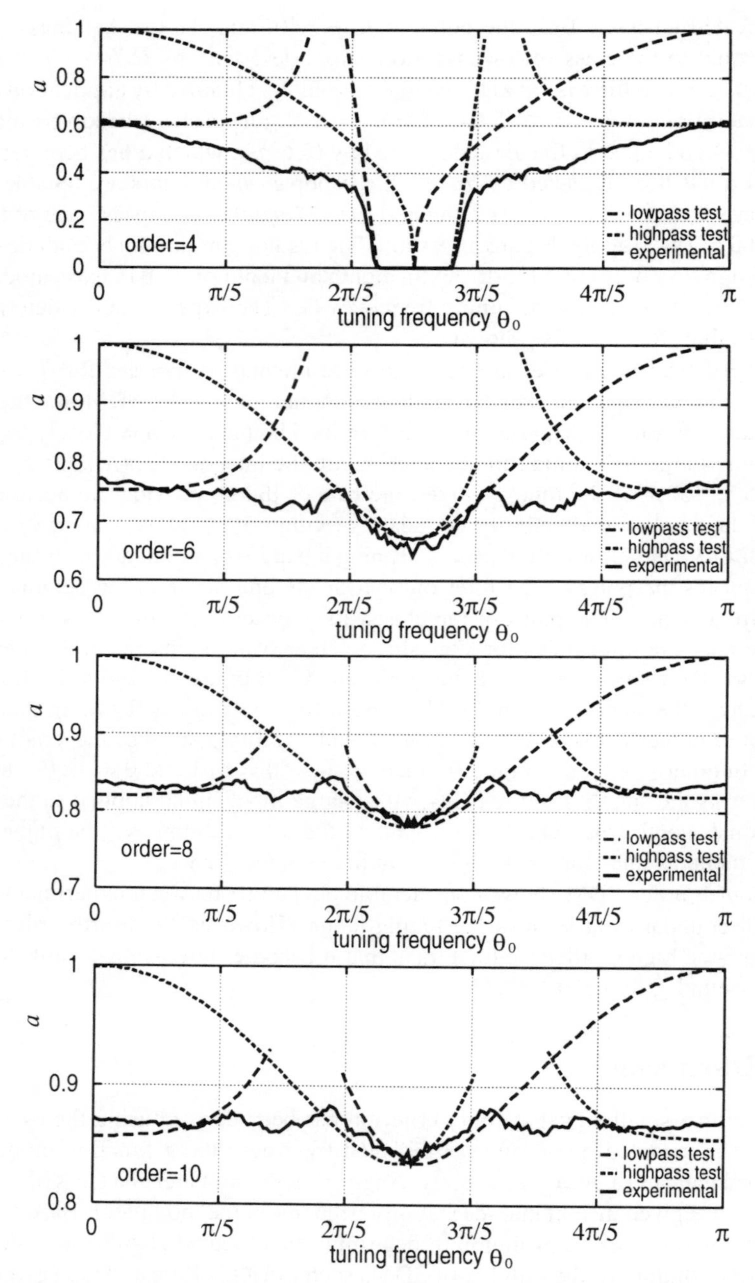

Figure 5.28: *Theoretically and experimentally determined minimum small-signal stability values of a as a function of the tuning frequency θ_0 of a fourth, sixth, eighth and tenth order bandpass SDM with a loop filter described by (5.70) with $b = 1$.*

stability boundary on a loop filter parameter for 60% of the tuning range between 0 and $f_s/2$. For the other 40% of the tuning range and for a fourth order bandpass modulator, the rule of thumb did not give acceptable results. The errors in the prediction of the small-signal stability may be attributed to a number of assumptions made in the analysis:

- The rule of thumb was derived for a class of lowpass modulators.
- The rule of thumb is based on root locus analysis using an approximation of the maximum phase uncertainty of a sampled quantizer.
- The movement of the closed-loop poles in the root locus is assumed to be continuous. As the modulator is a discrete time system, this assumption will be only valid to a certain extend if the amplitude of the signal inside the loop will not change significantly within a single sample period.
- The df method only models the signal transfer of the fundamental frequency. The effect of higher harmonics is neglected.

5.7 Large Signal Stability

Even though an SDM is designed to be small-signal stable using the criteria of the previous section, the modulator can still become unstable when an input signal is applied. An SDM is said to be *large signal stable* when the modulator is free of large-signal limit cycles and the states of the modulator remain bounded for a large 'in-band' input signal. The large-signal stability depends on the amplitude of the 'in-band' input signal. As was mentioned in the previous section, the effect of the input signal on the stability of the modulator can be modeled by a change of the quantizer model parameter λ. An increasing input amplitude results in an increasing amplitude in the input signal of the quantizer and consequently a decrease of λ. The smallest value of λ for which all the poles of the closed-loop system lie within the unit circle determines the maximum input amplitude for which the modulator will be stable.

5.7.1 Analysis

The minimum value of λ for which the roots remain within the unit circle can be determined by calculating λ_0 at the intersection of the root locus with the unit circle corresponding to an unstable large-signal limit cycle. In Fig. 5.29 an example of a root locus is shown of a third order lowpass modulator in the case that the phase uncertainty is not taken into account. The loop filter of the modulator is given by:

$$G(z) = \frac{(1-0.5z^{-1})^3}{(1-z^{-1})^3} - 1 \qquad (5.71)$$

The intersection of the root locus with the unit circle giving the minimum value for λ is indicated by an arrow. The corresponding value for λ equals $\lambda_0 = 0.2313$. Incorporating the phase uncertainty results in a whole set of intersections with the unit circle,

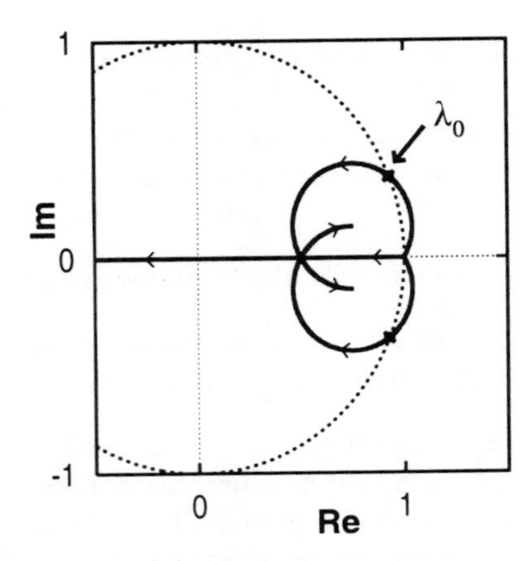

Figure 5.29: Root locus ($\alpha = 0$) of third order lowpass modulator with a loop filter according to (5.71).

as shown in Fig. 5.30. The set of intersections will give a range of values for λ for which the poles are on the unit circle. For the third order SDM of (5.71) this is range is given by: $0.1715 \leq \lambda_0 \leq 0.3480$. A worst-case value for the minimum λ providing

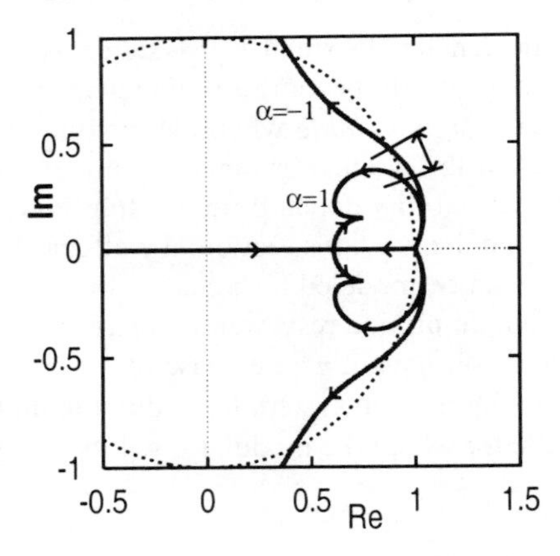

Figure 5.30: Range of root locus intersect points with unit circle resulting from the phase uncertainty.

large-signal stability can be obtained by taking the maximum value of this range. In other words, the minimum value for λ for which the poles of the system lie within the unit circle for any value of the phase uncertainty parameter $\alpha \in [-1, 1]$.

The resulting minimum value for λ will generally be more strict than the value determined without taking the phase uncertainty into account (i.e. $\alpha = 0$) as is done in [31], [51], and [29]. In [30, p.93] STIKVOORT already stated that a phase shift caused by the nonlinear part of the loop required the actual value for a stabilizing limiter in a noise shaper to be lower than the value found using the root locus method.

The worst-case value for the third order SDM equals $\lambda_0 = 0.3480$. Clearly, the phase uncertainty also a has a considerable impact on the large signal stability. The worst-case value of λ_0 has changed by more than 50% compared to the value determined from Fig. 5.29 without the phase uncertainty.

The relationship between the global signal gain λ and the input signal amplitude has been investigated thoroughly in [31]. As the quantizer model including the phase uncertainty only extends the linear gain model, the findings in [31] can be applied by simply using the more strict minimum value for the quantizer gain λ.

5.7.2 Stabilization Techniques

In order to guarantee the large-signal stability of a sigma delta modulator or noise shaper, several nonlinear stabilization techniques have been developed. One of these techniques is based on limiting of the signals inside the loop. According to the root locus analysis of SDMs, the quantizer gain λ should be kept above a certain minimum value λ_0. By limiting the states of the loop filter, the amplitude input signal of the quantizer can be kept below a certain value, resulting in a minimum gain value. Note that merely placing a limiter between the filter output and quantizer input or scaling the output signal of the filter does not affect the behavior of the SDM: the signal gain λ from the *filter output* to the *quantizer output* can still have any value between 0 and ∞. The limiter should be placed such that the dynamic behavior of the loop filter changes when the limiter is active, i.e. the limiter should affect the states of the loop filter. The limiting values for the states of the loop filter can be chosen such that the limiting will occur in steps. When the input to the modulator increases, more states will become limited. Effectively, the order of the loop filter is reduced in steps resulting in a graceful degradation of the performance [59]. An example of a lowpass SDM using such a stabilization technique is shown in Fig. 5.31. The last two integrators of the fourth order loop filter are limited. In the case that both limiters are active, the loop filter is reduced to a second order loop filter, resulting in a large-signal stable SDM.

In the case of a noise-shaper, a limiter can be placed inside the direct feedback path from the input of the quantizer to the input of the loop filter (see Fig. 5.32), and the relationship between the global gain of the nonlinear part of the loop (quantizer and limiter) and the amplitude of such a limiter can be determined [29]. Limiting the

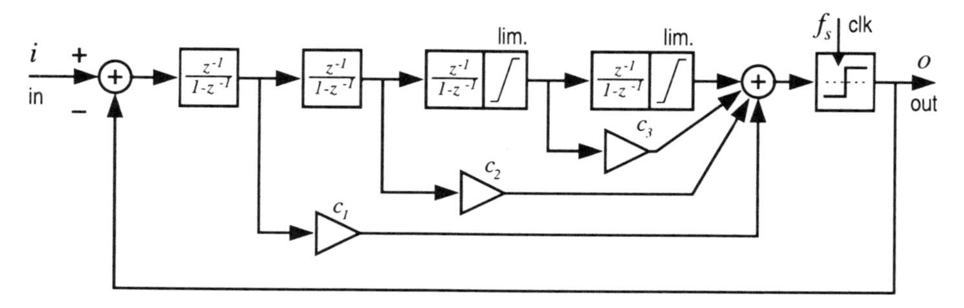

Figure 5.31: *Stabilization of a high order SDM by clipping of the state variables.*

states of the loop filter in one way or another will result in distortion in the output of the SDM and should be avoided under normal operating conditions of the modulator.

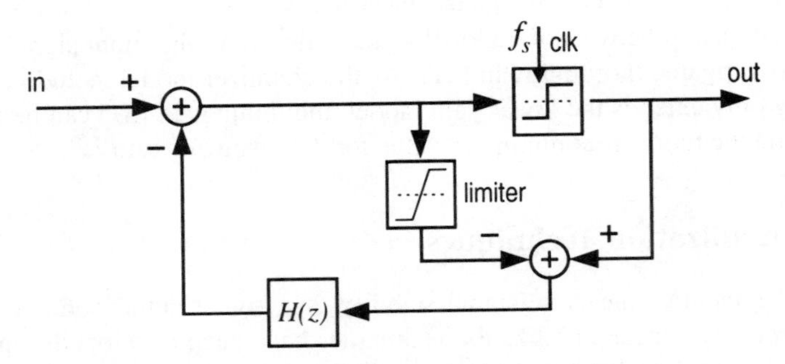

Figure 5.32: Stabilization of a noise shaper.

Another technique for obtaining stability of SDMs is resetting of filter states [2]. Large-signal limit cycles are detected by scanning the output of the modulator for suspicious patterns or by monitoring the amplitude of the signal at the input of the quantizer. When a suspicious pattern is found or the quantizer input amplitude exceeds a threshold, the loop filter states are reset to zero. Continuous overload of the modulator may cause periodic reset events and should also be avoided under normal operating conditions.

5.8 Relationship to the Noise Model

Comparison of the stability model of the SDM described in sec. 5.5 (see Fig. 5.33) and the performance prediction model used in chapter 4 (see Fig. 5.34) could suggest a relationship between the parameters of the two models. The parameter c_g of the

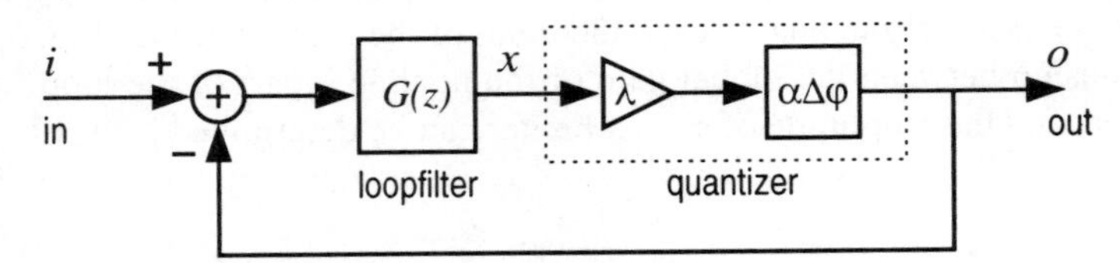

Figure 5.33: Stability model of an SDM.

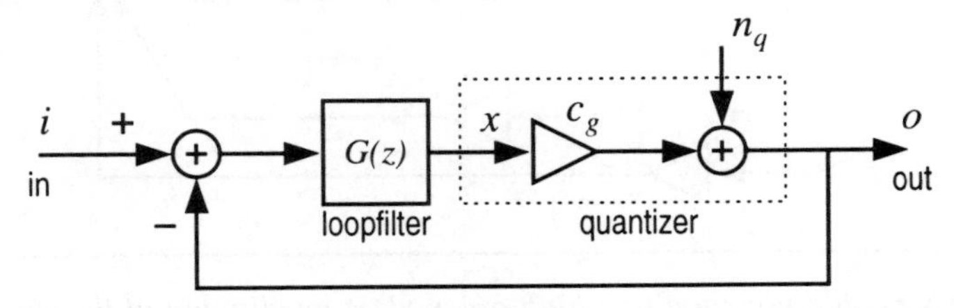

Figure 5.34: Linear performance prediction model of an SDM.

noise model and the parameters λ and α of the stability model represent the (complex) gain of the quantizer. However, the models are based on two different assumptions for the input signal of the quantizer. The parameter c_g represents the global signal gain when using the white noise and constant power assumptions for the quantization errors. The complex gain $\lambda \cdot e^{j\alpha\Delta\phi_{\max}(\theta)}$ is based on the assumption that the input signal of the quantizer is a sine wave. Nonetheless, a clear link between the two models is that the parameters result in the same singularities for the closed-loop transfer function as can be seen from eqs. (5.49) and (4.3). For the stability model, these singularities correspond to poles of the closed-loop transfer function. When these poles intersect with the unit circle at $z = e^{j\theta_i}$ for some λ and α, the stability model predicts a (stable or unstable) limit cycle. When $\alpha = 0$, $G(e^{j\theta_i})$ is real-valued. STIKVOORT showed in [30] that when $G(e^{j\theta_i})$ is real-valued, the noise model also exhibits a singularity for θ_i. The singularity in the noise model results in a peaking of the SDM output noise density at frequency θ_i for a certain value of c_g. As c_g depends on the input power of the SDM, the peaking of the output noise density depends on the power of the input signal. When the limit cycle with frequency θ_i is unstable, the peaking cannot be observed in practice as the modulator will be on the verge of instability. In practice, peaking of the output spectral density can be observed for stable limit cycles. For a small input amplitude, the output spectral density will exhibit a peak near the idle frequency as shown in Fig. 5.18. Unfortunately, this peak cannot be predicted by the noise model as the idle pattern corresponds to a singularity which occurs for a complex valued quantizer gain. Consequently, $G(e^{j\theta_i})$ will not be real-valued. The peaking of the output noise density can be demonstrated by replacing the real-valued c_g with the complex-valued quantizer gain from the stability model $\lambda \cdot e^{j\alpha\Delta\phi_{\max}(\theta)}$ and drawing the resulting predicted noise density. According to (4.4) the noise density at the output of the SDM equals:

$$N_o(\theta) = \frac{q^2}{24\pi} \cdot \left| \frac{1}{1 + c_g G(e^{j\theta})} \right|^2 \tag{5.72}$$

Substituting $c_g \to \lambda e^{j\alpha\Delta\phi_{\max}(\theta)}$ changes (5.72) into

$$N_o'(\theta) = \frac{q^2}{24\pi} \cdot \left| \frac{1}{1 + \lambda e^{j\alpha\Delta\phi_{\max}(\theta)} G(e^{j\theta})} \right|^2 \tag{5.73}$$

Figure 5.35 shows the resulting modified NTF of the second order lowpass modulator of sec. 5.3.1 for two sets of values for λ and α. For $\lambda = 0.887$ and $\alpha = -0.579$ the poles of the closed-loop system are at $z = \pm\, 0.98j$ and the modified NTF exhibits a peak at $f_s/4$. This situation resembles the actual output power spectral density at very low input amplitudes as shown in Fig. 5.18. When $\alpha = 0$ and λ is equal to the c_g resulting from the constant power requirement (4.6) for $P_i = 0$ ($\lambda = 0.667$), the resulting modified NTF resembles the actual output power spectral density for moderate input amplitudes as shown in Fig. 4.8.

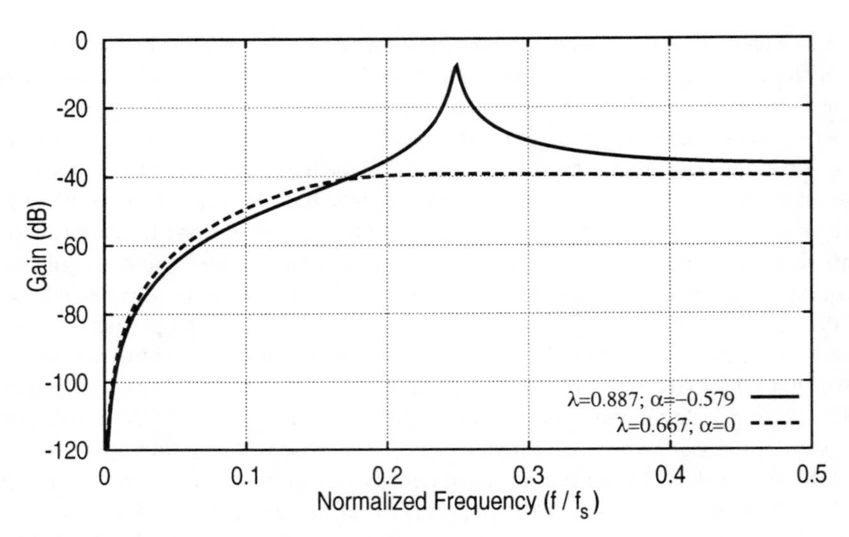

Figure 5.35: *Modified NTF according to (5.73) of a second order lowpass modulator.*

5.9 Conclusions

In this chapter the stability of sigma delta modulators has been investigated using the describing function method. The single gain model commonly used for the quantizer was shown to be inadequate for the prediction of all stability aspects of the behavior of the SDM.

Analysis showed that a sampled quantizer also exhibits a phase uncertainty. The phase uncertainty results from the inaccuracy in time with which quantizer-threshold crossings of the input signal are detected. In the case of a one-bit quantizer, the phase uncertainty solely depends on the frequency of the input sine wave. The phase uncertainty is zero for frequencies which are not a rational fraction of the sample frequency.

In order to improve the stability analysis, the linear gain model for the quantizer was extended with the phase uncertainty of the fundamental harmonic. An approximation for the phase uncertainty was developed to simplify the analytical expression of the phase uncertainty and the analysis of the root locus of the linearized system.

When using the extended model for the quantizer, the small signal stability boundaries for loop filter parameters of a class of lowpass SDMs were found. The theoretical boundaries lie within 5% of the experimental values. From the analysis of the root loci of these lowpass modulators a rule of thumb was derived for the small signal stability of lowpass and highpass modulators. In contrast to some other rules of thumb for the stability of these modulators, this rule of thumb is based on theoretical analysis instead of empirical results. Although this rule of thumb is neither necessary nor sufficient to guarantee small signal stability of an SDM, it can help in the design of practical SDMs. Sigma delta modulators usually achieve optimal (theoretical) performance when their design is near the small-signal stability boundary [30]. Designing an SDM with a larger small signal stability margin increases the maximum input amplitude for which the SDM is large-signal stable, but also reduces the in-band quantization error suppression.

Application of the rule of thumb to bandpass modulators gives varying results. For bandpass SDMs with a loop filter order higher than four, the rule of thumb gives an accurate estimate (within 5%) of the small signal stability boundaries on loop filter parameters for 60% of the tuning range between DC and $f_s/2$. The theoretical estimates for a fourth order bandpass SDM show a considerable deviation with the experimentally determined small signal stability boundaries. Although the rule of thumb derived for lowpass modulators can give reasonable results for bandpass modulators, care should taken when applying this rule of thumb to bandpass SDMs. Nonetheless, the analysis of the stability of the bandpass modulators shows that the phase uncertainty of the sampled quantizer has a considerable impact on the small signal stability properties of SDMs. The effect on the large signal stability is significantly lower but modeling of the phase uncertainty results in more accurate predictions of large-signal limit cycle amplitudes and the minimum stability value for the modeled quantizer gain.

Both the stability analysis incorporating the phase uncertainty and experimental results show that stability of non-chaotic modulators is mainly determined by the location of the zeroes of the loop filter transfer function and consequently the phase response of the loop filter. As the zeroes are located well inside the unit circle, the amplitude response of the loop filter is mainly determined by the poles which are located on or near the unit circle. Relocating the zeroes in the complex z-plane therefore mainly affects the phase response of the loop filter.

CHAPTER 6

DESIGN OF CONTINUOUS TIME BANDPASS SDMS

Bandpass sigma delta modulation is well suited for A/D conversion of narrow band signals modulated on a carrier signal. Although stability is an important design criterion for SDMs, several other criteria and performance goals determine the actual choice of architecture, oversampling ratio and other design considerations. For bandpass SDMs the use of a continuous time loop filter seems advantageous. The stability requirements on the equivalent discrete time loop filter affect the design and architecture of the continuous time loop filter.

6.1 Design Goals

As in the case of the design of many other analog circuits, the design of (integrated) sigma delta modulators is governed by several performance goals such as SNR and power consumption. In Table 6.1 the most important performance characteristics of an SDM are listed. In addition to these performance characteristics several other characteristics determine the overall merit of an SDM. The technology in which an integrated SDM is implemented may determine the applicability of the SDM in complete systems. A high output sample rate and a high number of parallel output bits of and SDM ADC may complicate the design of digital signal processing circuits following the SDM. Despite possible superior performance, the use of passive components is often unfavorable for economical reasons.

6.2 Design Considerations Overview

In section 3.6 a number of design considerations for SDM ADCs were mentioned. Most of these considerations affect the theoretically achievable (SNR) performance

Table 6.1: *Performance Characteristics of an SDM*

primary	**secondary**
Signal Bandwidth	Supply Voltage
Signal-to-Noise Ratio	Chip Area
Power Consumption	Input Sensitivity
Distortion	

of the modulator. A brief overview of the most important design considerations and trade-offs will be given here.

Architecture: The two most commonly used architectures are the single loop SDM and the multi stage (MASH) SDM. A multi stage SDM allows higher order noise shaping with loop filters of order one or two, thereby avoiding stability problems. The disadvantage of the MASH structure is its sensitivity to mismatch of quantizer levels and filter transfer functions.

Quantizer Resolution: Increasing the resolution of the quantizer in an SDM increases the theoretically achievable SNR of the modulator considerably. Unfortunately, the required accuracy of the implementation increases proportionally. The linearity of the DAC directly affects the performance of the SDM. A one bit DAC can achieve a high degree of linearity more easily than multi bit DACs as level mismatch of the two DAC output levels does not affect the linearity (see sec. 3.6.5).

Order of the Loop Filter(s): A high order loop filter results in a high theoretical SNR, even at low oversampling ratios. For a loop filter with an order larger than two, the stability of the SDM depends on the input signal and special measures have to be taken in order to avoid large-signal limit cycles.

Oversampling Ratio: Increasing the oversampling ratio is the easiest way for increasing the theoretical SNR of an SDM. Generally, changing the OSR does not affect the stability of the modulator nor does it require a more accurate implementation of the modulator. The major obstacle for very high sampling frequencies are the bandwidth requirements on the SDM circuitry itself and the signal processing circuitry following the SDM.

Discrete time vs. Continuous time Loop Filter(s): Discrete time (DT) loop filters can be implemented easily with great accuracy using Switched Capacitor (SC) techniques. The filter transfer function is not affected by the shape of the feedback signal. A key advantage of a continuous time (CT) loop filter is that the sampling operation on the input signal takes place inside the loop and sampling errors will also be shaped.

For bandpass modulators in particular, some additional considerations may affect the design of the SDM. First of all, a special architecture can be used. The complex SDM briefly described in section 3.3.3 allows the use of a lower order bandpass loop filter, thereby reducing the stability requirements on the design. A major disadvantage of these modulators is the required matching between the two channels representing the in-phase (I) and quadrature-phase (Q) component of the signal. Any mismatch results in cross-channel leakage when the I and Q signal components are recombined. This degrades the performance of the SDM.

Another consideration for bandpass modulators concerns the choice for a DT or CT loop filter. For an SC filter, the sample frequency f_s is limited by charge transfer accuracy requirements. By switching the circuit configuration of capacitors, electric charges are transfered from one capacitor to another. For reason of the limited bandwidth of opamps, the voltages of the nodes in a SC filter require a certain time to settle with the specified accuracy. Bandwidth limitations of opamps used in CT loop filters affect the overall transfer function of the loop filter but do not introduce additional distortion. Moreover, a CT bandpass loop filter can be tuned at a frequency other than $f_s/4$ without inferring additional hardware. In the case of a DT implementation, half of the filter coefficients is zero when the filter is tuned at $f_s/4$. This can be demonstrated easily by the following example. Using the transformation $z^{-1} \rightarrow -z^{-2}$ a second order lowpass (DT) filter $G_{\mathrm{lp}}(z)$ is changed into a bandpass filter $G_{\mathrm{bp}}(z)$ tuned at a quarter of the sample frequency $f_s/4$.

$$G_{\mathrm{lp}}(z) = \frac{2z^{-1} - z^{-2}}{1 - 2z^{-1} + z^{-2}} \xrightarrow{z^{-1} \rightarrow -z^{-2}} G_{\mathrm{bp}}(z) = \frac{-2z^{-2} - z^{-4}}{1 + 2z^{-2} + z^{-4}} \tag{6.1}$$

Writing the transfer of the bandpass loop filter in time domain gives

$$G_{\mathrm{bp}}: \quad x[k] = -2x[k-2] - x[k-4] - 2e[k-2] - e[k-4] \tag{6.2}$$

in which $x[k]$ is the filter output and $e[k]$ is the filter input signal. Clearly, the coefficients related to the input signal and output signal with an odd number of delays, e.g. $x[k-1]$ and $e[k-3]$, are zero and do not have to be implemented in the DT filter. An important reason for tuning the loop filter at a frequency other than $f_s/4$ is that aliasing of third order harmonics into the signal band is avoided. A third order harmonic distortion component of an input signal with frequency $f_s/4 + \Delta f$ aliases to the frequency $f_s/4 - 3\Delta f$ which falls into the signal band when Δf is small. Another source of distortion at $f_s/4$ is crosstalk of subharmonics of the clock frequency. The subharmonics will be most pronounced when digital signal processing circuitry is implemented on the same chip. Moving the tuning frequency to a frequency other than $f_s/4$ will reduce linearity requirements on the SDM circuitry slightly. Third order intermodulation products will still alias into the signal band, even when the tuning frequency is not equal to $f_s/4$. Crosstalk distortion can be reduced significantly by choosing the tuning frequency carefully.

6.3 Design Methodology

The first step in a design is the specification of the performance targets. Based on these targets several decisions are made with respect to the architecture of the SDM, the type of loop filter and the resolution of the quantizer. The remaining sections of this chapter will focus on the design of a single loop bandpass SDM with a continuous time loop filter and a one bit quantizer.

In order to achieve the specified performance targets (concerning SNR or DR and power consumption in particular), there is a trade-off between the order of the loop filter and the oversampling ratio. A higher order loop filter requires a lower oversampling ratio in order to achieve the same Dynamic Range (see Fig. 4.3). Although in general a high order loop filter consumes more power, a high oversampling ratio increases the the sample frequency and affects the power consumption of the digital signal processing circuitry following the SDM.

The design of an SDM with a continuous time loop filter will be done in discrete time domain. The sampled response of the CT loop filter to the pulse shape of the quantizer DAC can be replaced by an equivalent DT loop filter. The stability of the SDM is also analyzed in the discrete time domain. The DT loop filter should be designed such that the SDM is stable for small to moderate input signal amplitudes, without a significant reduction of the performance. Large signal stability can, for example, be achieved by clipping of filter state variables. Once the DT loop filter is designed, the CT loop filter transfer function can be determined from the DT loop filter and the pulse shape of the DAC. A direct transformation of the DT transfer function is possible using both time- and frequency-domain techniques [27,60]. For the resulting CT transfer function a suitable filter structure should be found. In the case of high order (> 4) loop filters this is generally a difficult task. Starting from a known CT filter structure the coefficients can be determined easily from the (reverse) CT to DT transformation. For this reverse method to succeed, the CT filter structure should have sufficient degrees of freedom to implement the required DT transfer function. When the coefficients of the CT filter are found, a suitable implementation of the CT loop filter can be designed.

From the above discussion a rough design methodology for single loop, one bit (bandpass) SDMs with a continuous time loop filter can be extracted:

1. Define the design goals (such as tuning frequency) and performance targets (i.e. SNR, power consumption and bandwidth).

2. Determine the oversampling ratio and order of the loop filter using theoretical estimates of the achievable dynamic range and possible power consumption limitations on the sample frequency.

3. Design the DT loop filter such that the SDM is small-signal stable. The poles of the loop filter determine the in-band gain and tuning frequency. The zeroes of the loop filter determine the stability of the modulator. An estimate for the small signal stability boundary for the loop filter parameter(s) can be found using the method[1] of chapter 5. The optimal values of these parameters with respect to the performance are near the small signal stability boundary.

4. Determine the equivalent DT loop filter of a CT loop filter structure having sufficient degrees of freedom. Note that the pulse shape of the DAC can have a considerable influence on the equivalent DT loop filter transfer function.

5. Match the designed DT filter transfer function with the equivalent DT loop filter transfer function and solve the CT loop filter coefficients.

6. Determine a suitable implementation of the CT loop filter.

In the following sections special attention is given to the CT to DT transformation of the loop filter. Using the properties of this transformation, it will be shown that an SDM with a CT loop filter can be subsampled. Subsampling allows A/D conversion of very narrow band signals on high IF frequencies with a relatively low sample

[1] Note that this method is only accurate when the (angular) tuning frequency lies within $[0, \pi/5]$, $[2\pi/5, 3\pi/5]$ or $[4\pi/5, \pi]$.

frequency. This chapter is completed by the presentation of some CT bandpass filter structures which have sufficient degrees of freedom to result in a desired DT transfer function.

6.4 Continuous time to Discrete time Transformation

As was mentioned in section 3.6.4 an SDM with a CT loop filter has a DT counterpart. In order to determine the coefficients of the CT loop filter, the equivalent DT loop has to be calculated. The response of the equivalent DT loop filter is defined by the sampled response of the CT loop filter to the pulse shape of the DAC. Figure 6.1 shows a graphical representation of the effective DT impulse response of the DAC followed by the CT loop filter.

Sampling of a signal can be represented mathematically by multiplication with a sum of time-shifted Dirac-pulses (δ) [48]. The equivalent DT loop filter can be written as:

$$G_{eq}(z) = \mathcal{Z}\left\{\mathcal{L}^{-1}\{F(p)R(p)\}\cdot\left(\sum_{k=0}^{\infty}\delta(t-kT_s)\right)\right\} \tag{6.3}$$

in which $F(p)$ is the CT loop filter transfer function, $R(p)$ the Laplace-transform of the DAC pulse shape, T_s is the sample period, $\mathcal{Z}$ represents the z-transform and $\mathcal{L}^{-1}$ the inverse Laplace transform. As a multiplication in time-domain is equivalent to a convolution in frequency-domain, eq. (6.3) can be written as:

$$G_{eq}(z) = \mathcal{Z}\left\{\mathcal{L}^{-1}\left\{(F(p)R(p)) * \frac{1}{1-e^{-pT_s}}\right\}\right\} \tag{6.4}$$

Writing out the convolution and applying $z = e^{pT_s}$ gives

$$G_{eq}(z) = \frac{1}{2\pi j}\int_{c-j\infty}^{c+j\infty}\frac{F(s)R(s)}{1-e^{sT_s}e^{-pT_s}}ds\bigg|_{z=e^{pT_s}} \tag{6.5}$$

The integral is calculated over a line in the complex s-plane parallel to the imaginary axis and can be changed into a contour integral by extending it with an infinite radius

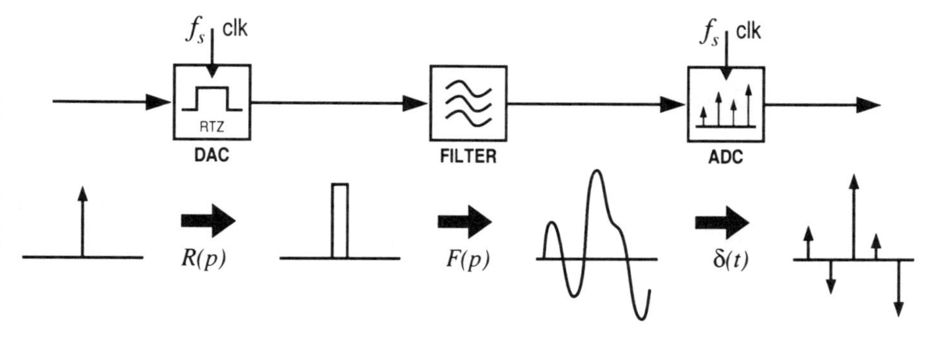

Figure 6.1: *Sampled response of the CT loop filter to the DAC pulse.*

semi-circle in the left half of the complex s-plane. Under the assumption that the integral over the semi-circle is zero, eq. (6.5) can be written as:

$$G_{eq}(z) = \frac{1}{2\pi j} \oint_{\Gamma} \frac{F(s)R(s)}{1 - e^{sT_s} e^{-pT_s}} ds \Bigg|_{z=e^{pT_s}} \tag{6.6}$$

in which Γ is the closed contour of an infinite radius semi-circle in the left hand side of the s-plane enclosing all singular points (poles) of $F(s)$ and $R(s)$ [48]. Using the residue theorem of CAUCHY, this integral can be solved [61]. The equivalent DT filter transfer function follows from:

$$G_{eq}(z) = \sum_{n=1}^{N} \operatorname*{Res}_{s=s_n} \frac{F(s)R(s)}{1 - e^{sT_s} z^{-1}} \tag{6.7}$$

in which s_n are the poles of $F(s)$ and $R(s)$. The residue $\operatorname{Res} f(x)$ for an mth order pole at $x = x_0$ is given by

$$\operatorname*{Res}_{x=x_0} f(x) = \frac{1}{(m-1)!} \lim_{x \to x_0} \left\{ \frac{d^{m-1}}{dx^{m-1}} [(x - x_0)^m f(x)] \right\} \tag{6.8}$$

As an example, the equivalent DT transfer function of a second order resonator will be calculated. The CT transfer function of a resonator equals:

$$F(p) = \frac{\omega_g (p + \omega_z)}{p^2 + p\omega_0/Q + \omega_0^2} \tag{6.9}$$

For the DAC an RTZ pulse shape with a duration of $\frac{1}{2}T_s$ is used. The impulse response $r(t)$ of the DAC pulse shape is given by:

$$r(t) = u(t - T_s/4) - u(t - 3T_s/4) \tag{6.10}$$

in which $u(t)$ is the unit step function: $u(t) = 0$ for $t < 0$ and $u(t) = 1$ for $t \geq 0$. The Laplace transform of the DAC pulse shape equals:

$$R(p) = \mathcal{L}\{r(t)\} = \frac{e^{-p\frac{T_s}{4}} - e^{-p\frac{3T_s}{4}}}{p} \tag{6.11}$$

The equivalent DT transfer function is calculated by eq. (6.6). Substituting $R(p)$ and $F(p)$ gives

$$G_{eq}(z) = \frac{1}{2\pi j} \oint_{\Gamma} \frac{(e^{-s\frac{T_s}{4}} - e^{-s\frac{3T_s}{4}})\omega_g(s + \omega_z)}{s(s^2 + s\omega_0/Q + \omega_0^2)(1 - e^{sT_s} z^{-1})} ds \tag{6.12}$$

Because of the terms $e^{-p\frac{T_s}{4}}$ and $e^{-p\frac{3T_s}{4}}$, the integrand does not converge in the left half of the complex plane. This problem is overcome by recognizing that e^{-sT_s} equals a delay of a full sample period and results in a term z^{-1} in the equivalent DT transfer function. Equation (6.12) can be rewritten as:

$$G_{eq}(z) = \frac{z^{-1}}{2\pi j} \oint_{\Gamma} \frac{(e^{s\frac{3T_s}{4}} - e^{s\frac{T_s}{4}})\omega_g(s + \omega_z)}{s(s^2 + s\omega_0/Q + \omega_0^2)(1 - e^{sT_s} z^{-1})} ds \tag{6.13}$$

As the integrand of (6.13) is convergent in the left half of the complex plane, the integral can be calculated by residues, giving:

$$G_{eq}(z) = z^{-1}\left(\left.\frac{(e^{s\frac{3T_s}{4}} - e^{s\frac{T_s}{4}})\omega_g(s+\omega_z)}{s(s-s_1^\star)(1-e^{sT_s}z^{-1})}\right|_{s\to s_1} + \left.\frac{(e^{s\frac{3T_s}{4}} - e^{s\frac{T_s}{4}})\omega_g(s+\omega_z)}{s(s-s_1)(1-e^{sT_s}z^{-1})}\right|_{s\to s_1^\star}\right) \tag{6.14}$$

with $s_1 = -\omega_0/2Q + j\omega_0\sqrt{1-1/4Q^2}$ and $s_1^\star$ the complex conjugate of s_1. Combining the two terms in (6.14) gives:

$$G_{eq}(z) = K_g \frac{z^{-1}(a_1 + a_2 z^{-1})}{1 - 2b\cos(\theta_0)z^{-1} + b^2 z^{-2}} \tag{6.15}$$

with for $Q \gg 1$:

$$\theta_0 \approx \omega_0 T_s \tag{6.16}$$

$$K_g \approx (\omega_g/\omega_0)\sin(\omega_0 T_s/4) \tag{6.17}$$

$$b = e^{-\theta_0/2Q} \tag{6.18}$$

$$a_1 \approx (\omega_z/\omega_0)\sin(\theta_0/2) - \cos(\theta_0/2) \tag{6.19}$$

$$a_2 \approx \cos(3\theta_0/2) - (\omega_z/\omega_0)\sin(3\theta_0/2) \tag{6.20}$$

Clearly, the poles of the equivalent DT transfer function are fully determined by the poles of the CT transfer function. The relationship between the CT pole s_1 and the DT pole z_1 is given by:

$$z_1 = e^{s_1 T_s} \tag{6.21}$$

The zeroes of the DT transfer function depend on both the poles and the zeroes of the CT filter transfer function. When the equivalent DT transfer function $G_{eq}(z)$ of the CT resonator has been determined, the coefficients ω_0, Q, ω_g and ω_z of the CT resonator can be found using eqs. (6.15)-(6.20). Note that the CT loop filter structure should have sufficient degrees of freedom to allow the implementation of the desired DT loop filter transfer function. For example, when ω_z is zero, the zero of the equivalent DT loop filter can not be chosen independently from θ_0.

Generally, the N poles of an Nth order DT loop filter are determined by the N poles of the CT loop filter. As an Nth order DT loop filter has at most $N-1$ zeroes, the CT loop filter should have $N-1$ degrees of freedom to set the location of the DT zeroes. This requirement sets limitations on the types of CT loop filters that can be used. In section 6.6 a number of bandpass loop filter structures is presented satisfying the aforementioned requirement.

With respect to the relationship between the CT and equivalent DT transfer function, it should be noted that solving eqs. (6.15)-(6.20) does not provide a unique solution for ω_0. It can be easily verified that for any value of ω_0 for which holds that

$$\omega_0 T_s = \pm\theta_0 + 2k\pi \quad \text{with:} \quad k \in \mathbb{Z} \tag{6.22}$$

an identical DT transfer function $G_{eq}(z)$ can be obtained by scaling the remaining coefficients ω_g, ω_z and Q. This means that an SDM with a CT loop filter can be subsampled. Substituting $T_s = 1/f_s$ and $\omega_0 = 2\pi f_0$ in (6.22) gives

$$f_0 = (\pm \tfrac{\theta_0}{2\pi} + k) f_s \quad \text{with:} \quad k \in \mathbb{Z} \tag{6.23}$$

Clearly, the tuning frequency f_0 can be larger than $f_s/2$. In that case, the frequency of the input signal should still be equal to f_0. This property can be used to design a subsampled continuous time modulator.

6.5 Subsampling in Continuous time SDMs

In [62] GOURGUE and BELLANGER introduced the concept of subsampling a continuous time bandpass SDM. The basic idea of subsampling is to undersample with respect to the center frequency of a narrow-band input signal and to oversample with respect to the bandwidth of the input signal. In Fig. 6.2 the basic scheme of a subsampled CT modulator is shown.

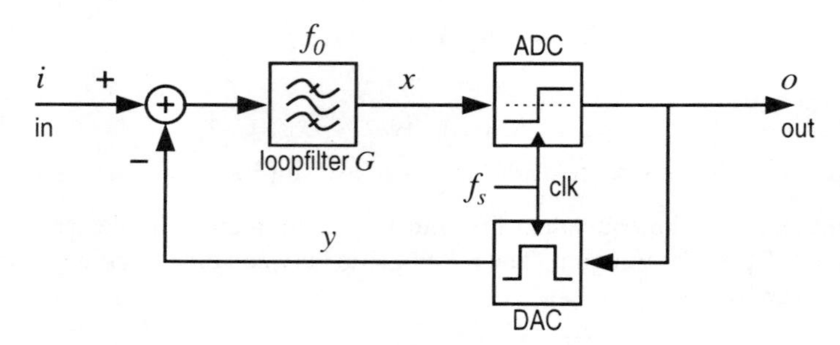

Figure 6.2: *Subsampled continuous time SDM ($f_0 > f_s/2$).*

The operation of the subsampled CT SDM can be explained by constructing the resulting spectra of the signals inside the SDM (see Fig. 6.3). The SDM is sampled at a rate f_s. The loop filter of the bandpass SDM is tuned to a frequency of $f_0 > f_s/2$. When an input signal with frequency f_0 is applied to the modulator, it will pass through the loop filter and will be sampled by the coarse ADC. As a result, the input signal will be aliased to a frequency f_a:

$$f_a = \pm(f_0 - K_{\text{ssf}} \cdot f_s) \tag{6.24}$$

with K_{ssf} an integer such that $f_a \in [0, f_s/2]$. The factor K_{ssf} will be called the subsample factor. The output spectrum will contain repeated versions of the input signal spectrum and the shaped quantization noise of the coarse quantizer. The output signal o is filtered by the hold-function of the DAC, resulting in the feedback signal y. In Fig. 6.3 an NRZ pulse shape is assumed for the DAC. In that case the feedback signal y is equal to the output signal o filtered by:

$$R(f) = T_s \frac{\sin(\pi f T_s)}{\pi f T_s} \tag{6.25}$$

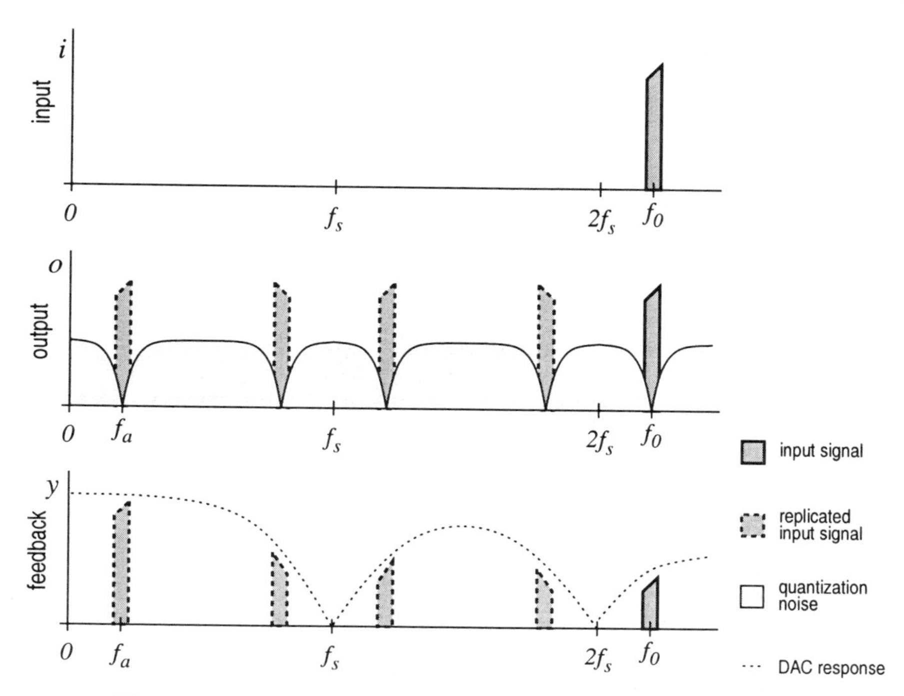

Figure 6.3: *Signal spectra in a subsampled continuous time SDM.*

The key to the operation of the subsampled CT SDM is that the DAC output spectrum repeats at multiples of the sample frequency. Unfortunately, the filtering operation by the hold function of the DAC severely affects the efficiency of the feedback loop. This is particularly true for high subsample factors. Theoretically, the loss of gain caused by the hold function can be compensated by the loop filter. However, the reduced output power of the DAC at f_0 causes a reduction of the maximum possible input signal amplitude. In the case that the (in-band) system noise exceeds the (in-band) quantization noise, the performance of the modulator depends on the absolute maximum input signal amplitude. A reduction of the maximum input amplitude will result in a reduction of the maximum SNR of the modulator.

GOURGUE and BELLANGER suggested to solve the loss of gain of the DAC at high frequencies by modulating (mixing) the feedback signal y with a carrier. By modulating the feedback signal, the large-amplitude replica of the input signal at f_a is upconverted to f_0, thereby increasing the efficiency of the DAC at the frequency f_0 and increasing the maximum possible input amplitude. As the mixer is placed within the feedback loop after the DAC, any additional distortion introduced by the mixer will add directly to the distortion in the output of the SDM.

Another solution for improving the efficiency of the DAC is a change of the pulse shape. Note that changing the pulse shape can be considered a special kind of mixing. In Fig. 6.4 four different pulse shapes are shown. The corresponding frequency responses are shown in Fig. 6.5. The four pulses are a Non-Return-to-Zero (NRZ) pulse, a Return-To-Zero pulse (RTZ), a double RTZ pulse (RTZ2) and a double com-

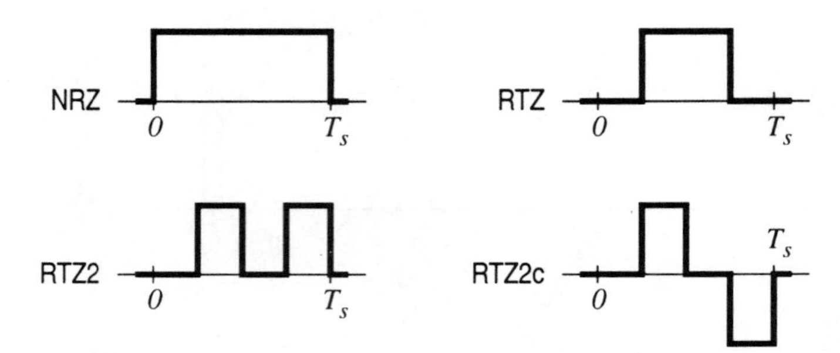

Figure 6.4: *Four different DAC pulse shapes.*

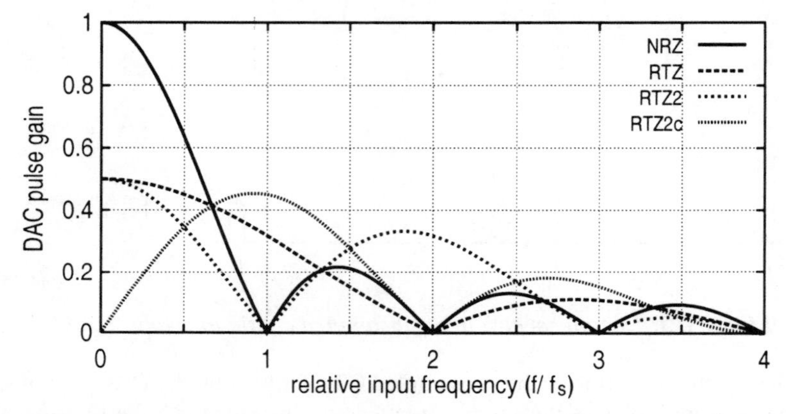

Figure 6.5: *Frequency response of the DAC pulse shapes shown in Fig. 6.4.*

plementary RTZ pulse (RTZ2c). The NRZ and RTZ pulses are optimized for signals near DC. For frequencies near f_s the RTZ2c pulse shows the largest gain. In the case that the signal frequencies are near $2f_s$ as Fig. 6.3, the RTZ2 pulse results in the smallest loss of gain.

Apart from the NRZ pulse, the four DAC pulses shown in Fig. 6.4 do not exhibit inter-symbol distortion due to limited rise and fall times. In order to be able to make the RTZ, RTZ2 and RTZ2c pulses, a clock frequency of $2f_s$ is required. Because no additional mixer is needed inside the feedback loop, the distortion caused by the DAC depends only on the accuracy of the implemented pulse shape.

In addition to the loss of gain, a high subsample factor causes additional design problems. For a given tuning frequency f_0, increasing the subsample factor will reduce the sample frequency f_s. When the bandwidth BW of the input signal is fixed, the oversampling ratio OSR will reduce when the subsample factor is increased. Note that a high oversampling ratio (OSR $\gg$ 100) combined with a high subsample factor ($K_{\mathrm{ssf}} \gg 1$) will result in practical design problems. In that case, the tuning frequency f_0 and the required bandwidth BW of the loop filter can be approximated by:

$$f_0 \approx K_{\mathrm{ssf}} \cdot f_s \quad \text{and:} \quad \mathrm{BW} = f_s/\mathrm{OSR} \tag{6.26}$$

As a result, the ratio between the tuning frequency and the bandwidth of the loop filter becomes very high. In the case of a second order bandpass filter this ratio is equal to the quality factor Q:

$$Q = \frac{f_0}{\text{BW}} \approx K_{\text{ssf}} \cdot \text{OSR} \gg 100 \tag{6.27}$$

Loop filters with such high quality factors are hard to design.

6.6 Bandpass Loop Filter Structures [63]

As was mentioned in sections 6.3 and 6.4, a CT loop filter structure should provide sufficient degrees of freedom to design the desired DT transfer function. A CT loop filter transfer function can be written as:

$$F(p) = K_g \frac{p^N + a_{N-1}p^{N-1} + a_{N-2}p^{N-2} + \cdots + a_1 p + a_0}{(p - p_{N-1}) \cdot (p - p_{N-2}) \cdots (p - p_1)(p - p_0)} \tag{6.28}$$

The N poles p_i of the CT loop filter determine the locations of the N poles of the DT loop filter. In order to place the $(N-1)$ DT zeroes independently, the coefficients $a_{N-1} \dots a_0$ should be independent. The contribution to the total loop filter transfer function of a coefficient a_i is approximately $a_i p^i / p^N$, which is a term of "order" $(N-i)$. As the coefficients a_i should be independent, the loop filter structure should contain N independent paths of different order.

Continuous time bandpass loop filters for SDMs can be designed using a cascade of resonators. Such a resonator may consist of a passive LC combination or active Gm-C sections. An active RC or Gm-C resonator has both in-phase (I) and quadrature-phase (Q) inputs and outputs. The relationship between the outputs and inputs is described in the Laplace-domain by:

$$\begin{aligned} \text{out}_I &= \omega_0 \frac{p \cdot \text{in}_Q + \omega_0 \cdot \text{in}_I}{p^2 + p\omega_0/Q + \omega_0^2} \\ \text{out}_Q &= \omega_0 \frac{p \cdot \text{in}_I - \omega_0 \cdot \text{in}_Q}{p^2 + p\omega_0/Q + \omega_0^2} \end{aligned} \tag{6.29}$$

For the in-phase output out_I, the in-phase input in_I has an effective order of two $(1/p^2)$ and the quadrature input in_Q has an effective order of one (p/p^2). For the quadrature-phase output out_Q, the effective orders are interchanged. Using these resonators, several CT bandpass filter structures can be designed having sufficient degrees of freedom to result in the desired DT transfer function. Figure 6.6 shows a bandpass filter based on a cascade of resonators with a distributed output. The output of the loop filter is formed by summing the weighed outputs of the resonators. The filter in Fig. 6.7 consists of a cascade of resonators with distributed inputs. In order to have sufficient degrees of freedom in these filter structures, the two inputs or the two outputs of a resonator have to be used. This requirement prevents the use of passive LC resonators with one input (current) and one output (voltage). The modified filter structure in

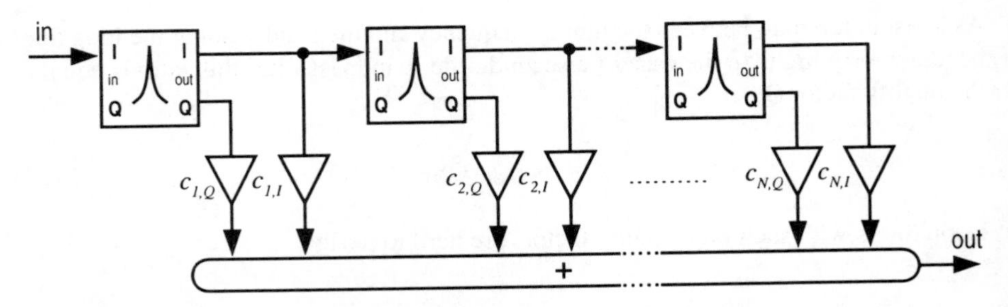

Figure 6.6: Cascade of resonators with distributed output.

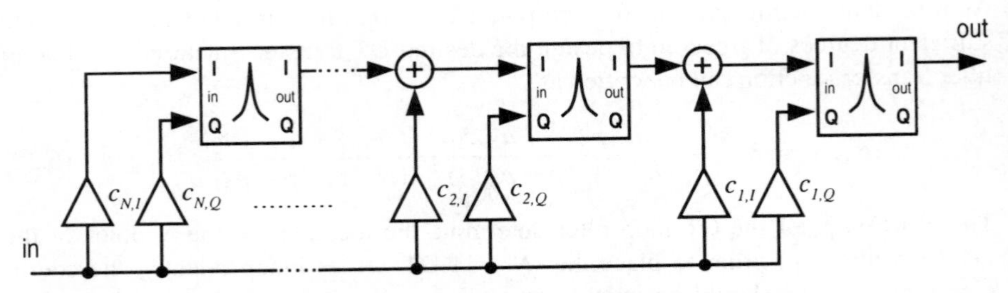

Figure 6.7: Cascade of resonators with distributed input.

Fig. 6.8 does allow the use of several LC resonators without reducing the degree of freedom. Using only one output of each resonator also simplifies the implementation of the limiters for large signal stability purposes. When the single output of the resonator is clipped, two feedforward paths are limited, thereby reducing the effective order of the loop filter by two. This helps to retain the bandpass characteristic under clipping conditions as the number of poles remains an even number.

Although the use of passive LC resonators is often frowned upon, it can improve the performance drastically. Using an LC resonator as the first resonator in a bandpass loop filter improves the noise figure of the filter considerably. Additionally, on-chip

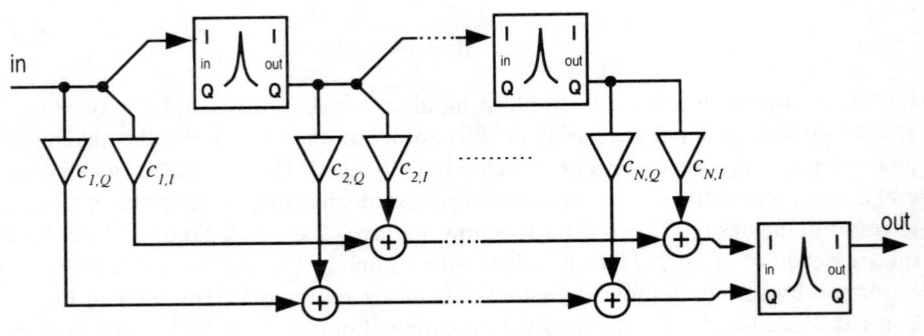

Figure 6.8: Cascade of resonators with combined distribution of input and output.

passive LC resonators might prove to be a useful solution for a single-chip bandpass SDM at radio frequencies (RF).

6.7 Conclusions

In this chapter a straightforward design method for continuous time SDMs is presented. The stability analysis and design of the loop filter are done in the discrete time domain. The design of the continuous time loop filter is based on continuous time to discrete time transformation. The CT filter structure used in the transformation should generate sufficient degrees of freedom for realizing the desired DT loop filter transfer function. A set of bandpass filter structures based on a cascade of resonators is proposed. The CT bandpass filter structures satisfy the requirement for sufficient degrees of freedom.

CHAPTER 7

SDM IMPLEMENTATIONS

During the research that lead to this book, several implementations of (bandpass) SDMs have been realized. A digital SDM board was made for real-time simulations. The design of two continuous time bandpass SDMs was based on the stability criteria described in Chapter 5 and the loop filter structures proposed in Chapter 6.

7.1 Digital Test Set-Up

The lack of a fundamental understanding of the behavior of sigma delta modulators has forced designers to make extensive use of simulations. Such a simulation can be done using a behavioral model or at a circuit level. In order to investigate the *stability* of the modulator, the simulations can be done at a behavioral level, using a discrete time model of the modulator. Despite the high level of these simulations, the total simulation time usually is large. This is due to the fact that many time steps are required for a reliable prognoses of the stability of the modulator. For a certain input signal, an SDM may appear to be stable for as many as 10^9 time steps before the states of the SDM grow out of bounds or the SDM enters a large signal limit cycle. Although the computational power of computers increases significantly from generation to generation, simulation of such a large number of time steps still requires a considerable time. For example, a simulation of 10^9 time steps of a sixth order modulator using double precision floats takes approximately 100 minutes on a 133MHz Intel Pentium™ processor. This corresponds to an effective sample frequency of $10^9/6000 \approx 167$kHz.

In order to reduce the simulation time, a digital, programmable hardware test set-up was designed. The core of the set-up consisted of a programmable hardware device (PLD) from Altera [64]. The PLD (Altera EPF81500) contains 1500 flip-flops and more than 15000 logical ports. The configuration is stored in SRAM and can be downloaded using a serial or parallel link with a personal computer (PC). The programming of the device can be done in VHDL. An example of a VHDL implementation of a digital SDM is listed in Appendix D. In addition to the PLD core, the test set-up contains a 12 bit ADC (Philips TDA 8768) front-end and a 10 bit DAC (Philips TDA 8776) back-end. Figure 7.1 shows a block diagram of the complete set-up.

The loop filter of the all-digital SDM was designed with a direct form structure as shown in Fig. 3.14. The states of the loop filter have a 28 bit fixed point internal representation. The feedback level of the quantizer was chosen to be equal to the (MSB-4)th bit as the filter states can become considerably larger than the quantizer level. As a result, the filter states have a "headroom" of 4 bit and an absolute accuracy of 24 bit. The filter coefficients were implemented using a "shift and add" technique with up to four adders per coefficient. The coefficients were realized using Canonical

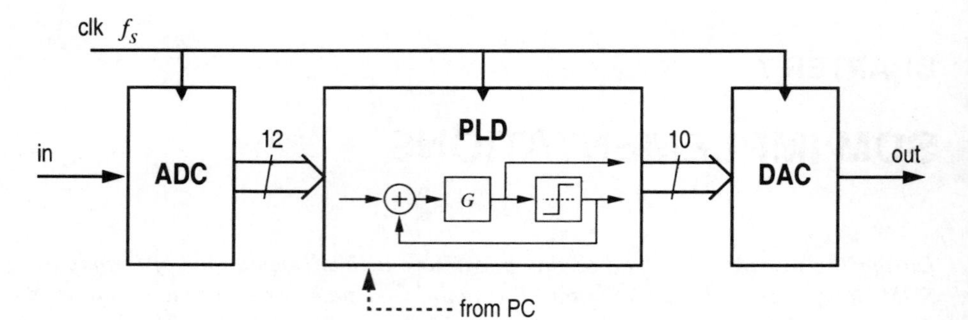

Figure 7.1: Hardware test set-up for real-time simulation of all-digital SDMs.

Signed Digit (CSD) coding. A CSD code is a trinary code in which each digit can have the value +1, 0 or -1. The value -1 is commonly represented by a 'T'. Similarly to the binary code, the value of the mth digit in a CSD code is multiplied by 2^m. By adding a sign to each digit of a binary code, the number of non-zero digits may be reduced. Table 7.1 shows a list of decimal values and the corresponding binary codes and CSD codes. As the hardware implementation of an 'add' operation is almost identical to the implementation of a 'subtract' operation, the number of adders needed for implementation of the coefficients can be reduced using CSD coding. It can be shown easily that CSD coded coefficients with up to four non-zero digits have at least 8 bit accuracy. In order to increase the amplitude range of the input signal, signal scaling was achieved by (internally) shifting the 12 bit ADC signal within the 28 bit signal representation of the all digital SDM.

This hardware set-up allows for real-time and long term evaluation of the behavior of digital SDMs. From the manufacturer-supplied timing diagrams, a worst-case maximum sample frequency of 2 MHz was determined for an SDM with 28 bit loop filter state representation and four adders per filter coefficient. The true maximum sample rate of the digital SDM depends on the actual loop filter coefficients. Sample rates up to 10MHz have been achieved with this hardware set-up. These sample rates clearly exceed the effective sample rate of software simulation.

The digital implementation of the SDM allows exact control of the loop filter coefficients. The quantized nature of the coefficients limits the coefficient resolution, but does not affect the behavior of the SDM. The fixed point implementation of the filter states introduces truncation and loss of precision. These errors may result in inter-

Table 7.1: Decimal numbers with corresponding binary and CSD codes.

Decimal	Binary	CSD
13	1101	1101
14	1110	100T0
29	11101	100T01
1.125	1.001	1.001
2.75	10.11	10.11
3.75	11.11	100.0T

nal limit cycles within the loop filter and reduce the noise suppression of the SDM. Such effects were only observed under extreme low input signal amplitude conditions combined with ideal integrators or resonators in the loop filter of the SDM.

The analog front-end and back-end of the test set-up simplifies the control of input signals and evaluation of output signals. Because of the ADC, analog signal generators can be used. Frequency modulated and time-varying signals can be easily applied to the modulator. The DAC allows real-time evaluation of the quantizer output and filter state signals in both frequency and time domain. The DAC eliminates the need of a high-speed FFT analyzer. Although the resolution of the DAC is limited, it is more than sufficient to examine the stability properties of the digital SDM. In the case that the one-bit quantizer output is monitored, not the resolution but the distortion on the pulse shape of the DAC determines the effective resolution with which the SDM performance can be measured. By connecting the quantizer output to the MSB of the DAC, the distortion can be minimized.

7.2 Discrete Fourth Order bandpass SDM [65]

In this section the design and implementation of a discrete fourth order CT bandpass SDM is presented. The design of the SDM was done using the methodology described in section 6.3.

7.2.1 Application

A bandpass modulator is well suited for A/D conversion of narrow band signals modulated on a carrier. The requirement of a high OSR can be easily satisfied as the signal bandwidth BW is usually very small compared to the center frequency f_c. A good example of such narrow band signals are AM and FM radio signals. The bandwidth[1] of AM and FM signals is 9kHz and 200kHz respectively. In radio receivers, these signals are modulated on an intermediate frequency (IF) of 10.7MHz.

A bandpass SDM can be used to digitize the IF modulated signals in a AM/FM receiver. This moves the IF signal processing stage to the digital domain. In Fig. 7.2 a block diagram of a receiver using a bandpass SDM is shown together with a block diagram of a modern analog (AM) receiver [66]. The analog receiver consists of a Low Noise Amplifier (LNA) followed by a bandpass filter (BPF) and a mixer converting the signals to the IF frequency of 10.7MHz using a tunable local oscillator (LO). The mixer is followed by a ceramic filter. A second mixer converts the signals to a second IF of 450kHz. A second ceramic channel selection filter precedes an automatic gain control (AGC) amplifier and the AM detection (or demodulation) stage. An audio power amplifier (P.A.) and loudspeaker complete the total receiver. The first stage of a receiver using IF A/D conversion is identical to the analog receiver. After the first ceramic filter, the IF signal is digitized and processed using a digital signal processing (DSP) unit. Additional filtering, final channel selection and demodulation of the (AM) signals is performed by the DSP unit. The audio output signal of the DSP unit is fed to

[1]Here, the term bandwidth refers to the commonly allocated frequency bandwidth or 'channel spacing'.

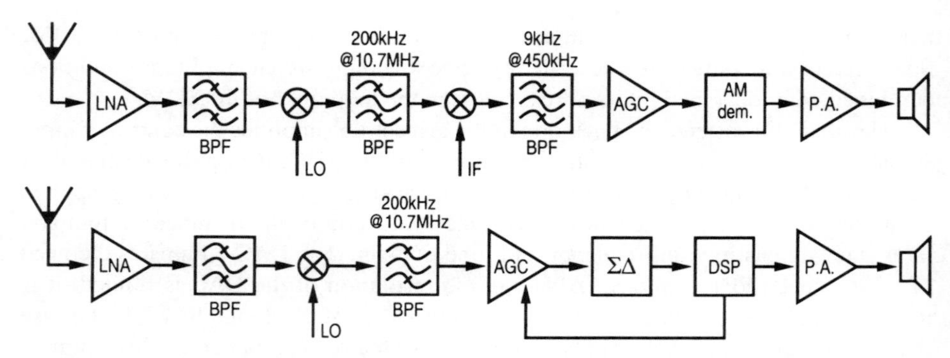

Figure 7.2: *Block diagram of a modern analog (AM) receiver (top) and a receiver using a bandpass SDM (bottom).*

the P.A. followed by the loudspeaker. Note that the receiver containing the bandpass SDM can also be used to receive FM signals by merely changing the DSP functionality.

Important specifications for communication building blocks are the SNDR and the intermodulation IM3. In communication systems using frequency division multiplexing such as AM and FM radio the intermodulation specifies the amount of cross-talk of adjacent channels.

7.2.2 Discrete Time Filter Design

As mentioned in chapter 6, the design of the loop filter of a continuous time bandpass SDM can be done in the discrete time domain. The loop filter is chosen from the class described by (5.70). The loop filter transfer function can be written as:

$$G(z) = \frac{(1+ae^{j\theta_0}z^{-1})^2(1+ae^{-j\theta_0}z^{-1})^2}{(1+be^{j\theta_0}z^{-1})^2(1+be^{-j\theta_0}z^{-1})^2} - 1 \tag{7.1}$$

Although for an optimal performance the complex conjugate poles should be spread within the signal frequency bandwidth, placing both pairs at the same frequency simplifies the design at the expense of of 3.5dB in DR. The order of the loop filter is four. The performance of a fourth order bandpass SDM is approximately equal to the performance of a second order lowpass SDM. In order to have a DR of at least 80dB, the OSR should be at least OSR=100 in the ideal case that $b = 1$ (see Fig. 4.3). Combined with a desired bandwidth of 200kHz this results in a sample frequency of $f_s = 40$MHz. The absolute tuning frequency equals $f_0 = 10.7$MHz, giving an angular tuning frequency of $\theta_0 = 1.681$ rad. According to Fig. 5.28, the parameter a which determines the location of the loop filter zeroes can be chosen $a = 0$. This value places the resulting NTF poles in the origin of the complex plane, thereby optimizing the performance. For $a = 0$ and $\theta_0 = 1.681$ rad, the fourth order bandpass SDM is large signal stable. Ideally, the poles of the loop filter lie on the unit circle $b = 1$. In reality, the quality factor will not be infinitely large. For a quality factor $Q = 40$, the pole radius equals $b \approx 0.98$ and the DT loop filter transfer function equals:

$$G(z) = \frac{d_1 z^{-1} + d_2 z^{-2} + d_3 z^{-3} + d_4 z^4}{(1 + b e^{j\theta_0} z^{-1})^2 (1 + b e^{-j\theta_0} z^{-1})^2} \tag{7.2}$$

with:

$$\begin{array}{ll} d_1 = -0.4302 & d_3 = -0.4131 \\ d_2 = -1.9671 & d_4 = -0.9224 \\ b = 0.98 & \theta_0 = 1.681 \end{array} \tag{7.3}$$

A plot of the resulting Noise Transfer Function $\mathrm{NTF}(z) = 1/(1+G(z))$ is shown in Fig. 7.3. The noise suppression at $f/f_s = 10.7/40 = 0.2675$ is clearly visible. Note that the out-of-band gain of the NTF exceeds 15dB. According to LEE's stability rule ($|\mathrm{NTF}(e^{j\theta})|^2 < 2$; see sec. 5.2), this modulator would be unstable. However, both simulations and the practical implementation did not show unstable behavior such as large-signal limit cycles.

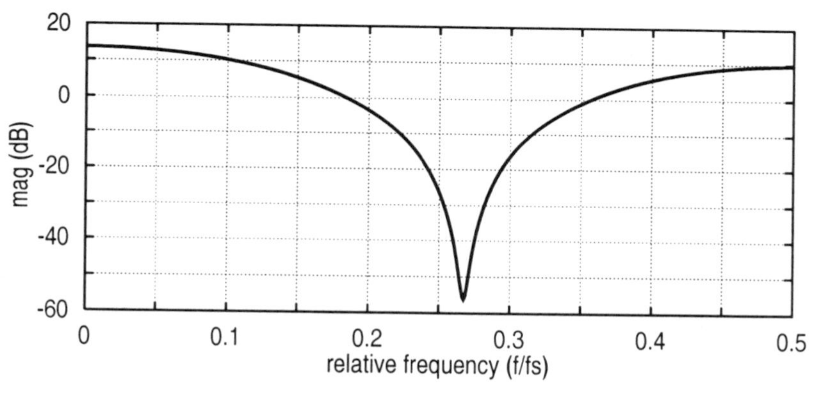

Figure 7.3: NTF of fourth order bandpass filter.

The maximum SNR follows from the linear prediction method described in sec. 4.1. The maximum SNR of 68dB in a bandwidth of 200kHz occurs for an input power of -6dB relative to the quantizer output power. Simulations show that the maximum SNR indeed is 68dB. A typical (simulated) output spectrum of the SDM is shown in Fig. 7.4.

7.2.3 Continuous Time Filter Design

When the DT loop filter transfer function $G(z)$ has been designed, the continuous time loop filter $F(p)$ can be calculated, taking into account the pulse shape of the DAC. A general fourth order CT bandpass filter transfer function with conjugate poles at f_0 can be written as:

$$F(p) = K_g \frac{p^3 + a_2 p^2 + a_1 p + a_0}{(p^2 + p\omega_0/Q_1 + \omega_0^2)(p^2 + p\omega_0/Q_2 + \omega_0^2)} \tag{7.4}$$

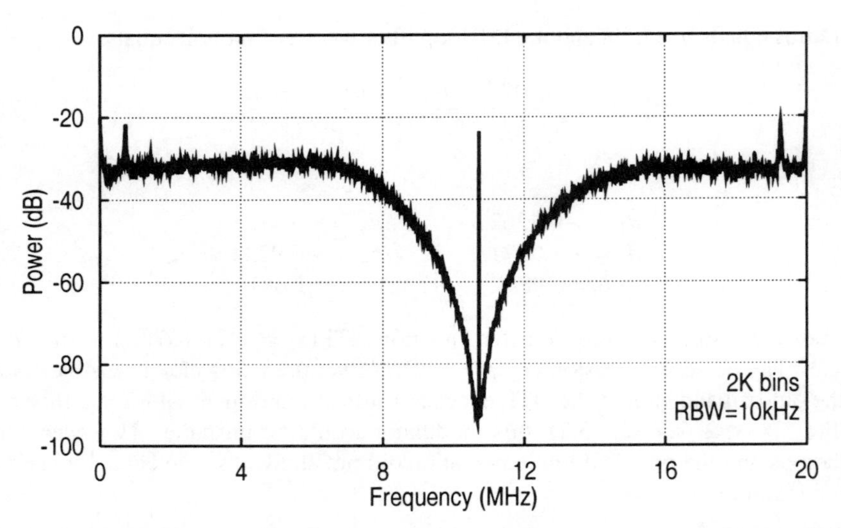

Figure 7.4: *Typical output spectrum (simulation) of a fourth order SDM with loop filter (7.2). Input power is -23dB relative to the quantizer output power.*

As the three zeroes of the DT transfer function were placed independently, at least three degrees of freedom are required in the CT loop filter transfer function (see sec. 6.4). As the poles are determined by the poles of the DT loop filter, all three the coefficients of the numerator of (7.4) have to be independent. A bandpass filter structure that consists of two high-Q LC resonators [67] as shown in Fig. 7.5 has one degree of freedom ($g_1 g_2 / g_3$) to place the zeroes of the loop filter. The second degree of freedom

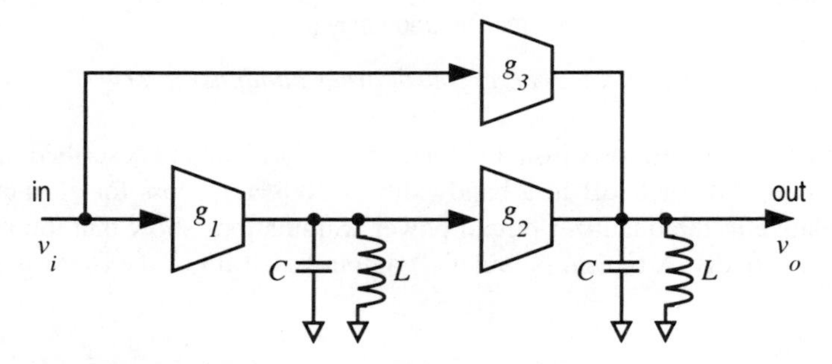

Figure 7.5: *A fourth order bandpass filter using two LC resonators.*

gives the overall gain of the filter. As the loop filter is placed before a one bit quantizer, the absolute gain of the loop filter does not affect the behavior of the SDM. As a result, the filter structure shown in Fig. 7.5 cannot be used for realization of the desired DT transfer function.

Several loop filter structures were proposed in section 6.6. The resonators in these filter structures should have both in-phase (I) and quadrature-phase (Q) inputs or I and Q outputs. This inhibits the use of an LC resonator to realize a high Q. The resonators in such filters should be realized using transconductance-capacitance (gmC) resonators. Only the filter structure of Fig. 6.8 allows the use of an LC resonator and ensures sufficient degrees of freedom to realize the desired DT transfer function. The resulting structure of the fourth order bandpass loop filter is shown in Fig. 7.6. For

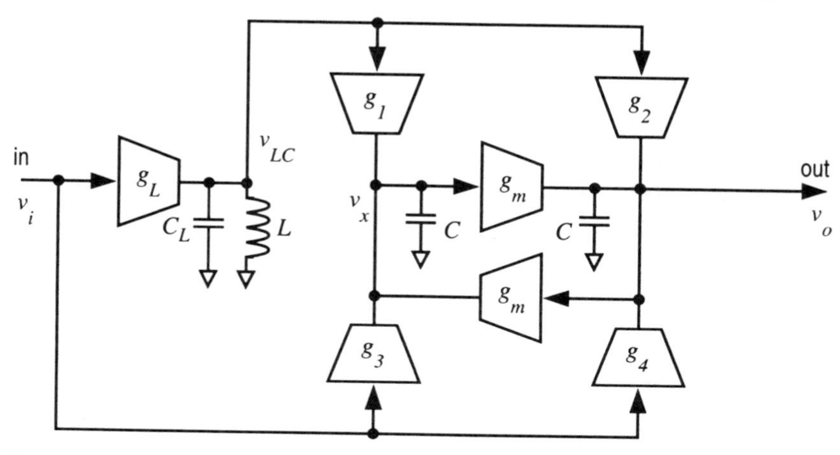

Figure 7.6: *A fourth order bandpass filter using an LC resonator combined with a gmC resonator.*

high values of the quality factors $Q_{\rm LC}$ and $Q_{\rm gmC}$, the transfer function of the filter in Fig. 7.6 is given by:

$$F(p) = \frac{g_4}{C}\frac{p^3 + \omega_c p^2(c_2\frac{\omega_L}{\omega_0} + c_3) + \omega_0^2 p(1 + c_1\frac{\omega_L}{\omega_0}) + \omega_0^3 c_3}{(p^2 + p\frac{\omega_0}{Q_{\rm LC}} + \omega_0^2)(p^2 + p\frac{\omega_0}{Q_{\rm gmC}} + \omega_0^2)} \tag{7.5}$$

in which $\omega_L = g_L/C_L$, $g_m/C = \sqrt{1/LC_L} = \omega_0$ and $c_{\rm i} = g_{\rm i}/g_4$ for i $= 1\ldots3$. It can be easily shown that the three coefficients of the numerator can be set independently. Using the CT to DT transformation described in section 6.4, the equivalent DT transfer function $G_{eq}(z)$ can be calculated:

$$G_{eq}(z) = \frac{1}{2\pi j}\int_{c-j\infty}^{c+j\infty}\frac{F(s)R(s)}{1 - e^{sT_s}e^{-pT_s}}ds\,\bigg|_{z=e^{pT_s}} \tag{7.6}$$

The CT to DT transformation takes into account the pulse shape $R(p)$ of the DAC. Here, a RTZ pulse with a duration of $T_s/2$ is used (see Fig. 7.7). The Laplace transform is given by:

$$R(p) = \frac{e^{-pT_s/4} - e^{-3pT_s/4}}{p} \tag{7.7}$$

The CT to DT transformation of eqs. (7.6), (7.5) and (7.7) is solved with the help of Maple [68], a mathematics software tool. The equivalent DT transfer function is then

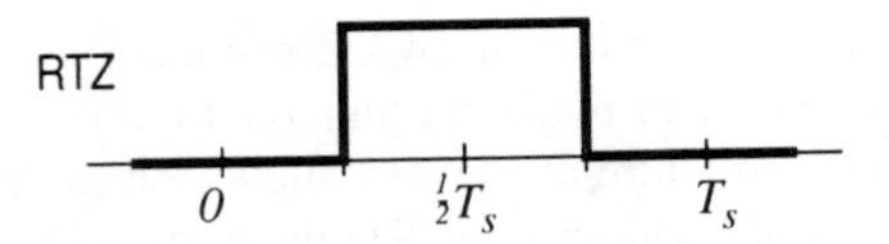

Figure 7.7: *A Return-to-Zero pulse with a duration of $T_s/2$.*

matched with the designed DT transfer function from eq. (7.2):

$$G_{eq}(z) = G(z) \tag{7.8}$$

From (7.8), the coefficients c_1, c_2 and c_3 can be solved by substituting all known numerical values and choosing $\omega_L = \omega_0$ and $g_4/C = \omega_0$:

$$\begin{aligned} c_1 &= -0.8167 \\ c_2 &= 0.4650 \\ c_3 &= -1.8054 \end{aligned} \tag{7.9}$$

7.2.4 Implementation

The CT loop filter of Fig. 7.6 consists of an LC resonator and a gmC resonator section. The implementation details are shown in Fig. 7.8. The LC resonator was implemented with an off-the-shelf IF bandpass filter (transformer), commonly used for IF filtering in AM/FM receivers. The corresponding transconductance g_L was implemented with a degenerated differential pair (Q1, Q2 and R_g). The quality factor of the LC resonator

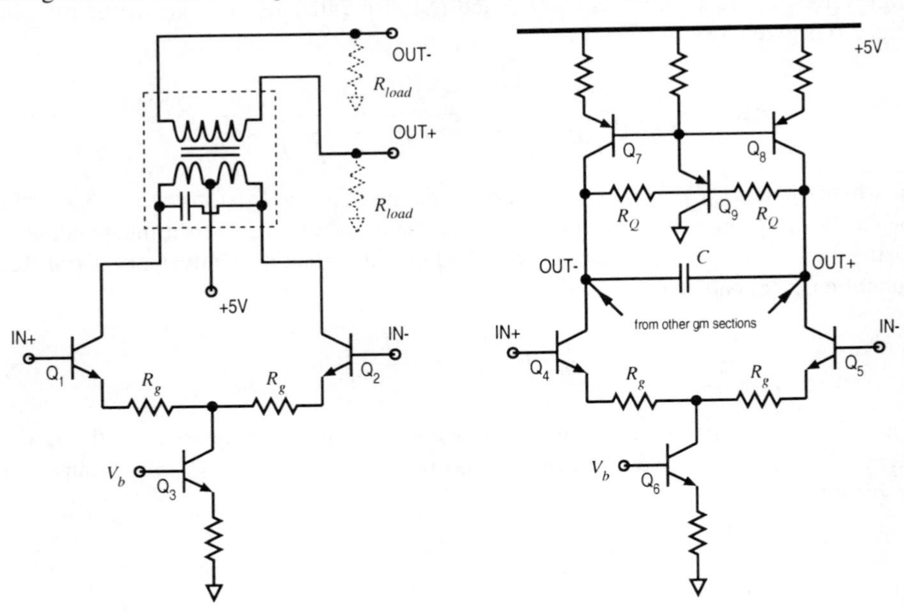

Figure 7.8: *Implementation details of fourth order bandpass filter: an LC resonator with transconductance (left) and a gmC stage (right).*

is given by the quality factor of the transformer and the load impedance R_{load}. The effective quality factor equals:

$$Q_{\mathrm{LC}} = \frac{Q}{1 + Q^2 \frac{R_L}{R_{load}}} \tag{7.10}$$

in which R_L is the parasitic resistance of the inductors of the transformer, and Q is the quality factor of the transformer $Q = \sqrt{L/CR_L^2}$. The gmC stages were also implemented differentially. Again a degenerated differential pair (Q4, Q5 and R_g) is used for the transconductance. The current-sources with common mode feedback (Q7, Q8 and Q9) determine the DC bias point. The quality factor Q_{gmC} of the gmC resonator is determined by the collector impedance of Q4 and Q5, in parallel to the resistors R_Q. As the collector impedance of Q4 and Q5 will generally be much larger than R_Q, the quality factor Q_{gmC} is mainly determined by R_Q. A custom-made bipolar one bit ADC and DAC was used to implement the quantizer. The ADC consisted of two flip-flops in master-slave configuration to decrease the decision time and reduce the bit error rate.

A test circuit was build on a bread-board using discrete devices to verify the design. The total power of the test circuit was 440mW. A considerable amount of the power is consumed by the one bit ADC and DAC which used ECL in- and output gates. The filter consumed 275mW. The large power consumption of the filter is caused by the required DC bias currents of 1mA for the discrete bipolar transistors.

7.2.5 Measurements

The quality factor Q_{gmC} of the gmC section could not be made very large in the bread-board implementation. Figure 7.9 shows the calculated and measured filter transfer function. The deviation in the measured transfer function consists mainly of a loss of gain and a parasitic linear phase (delay). The -6dB points of the filter are at -28kHz and +35kHz offset from the center frequency.

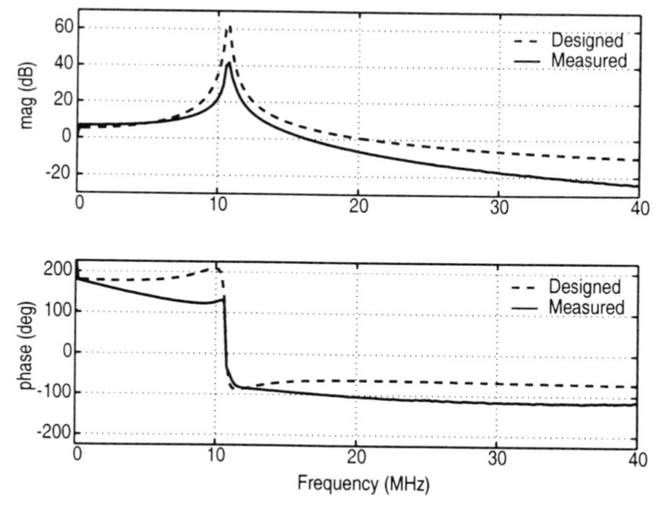

Figure 7.9: Designed and measured loop filter transfer characteristic.

A typical output spectrum of the fourth order SDM is shown in Fig. 7.10; the in-band region of the output spectrum is shown in Fig. 7.11. The output spectrum shows a large bump in the quantization noise near 13.3MHz. This suggests that the realized SDM prefers coding of the quantization noise with an idle frequency of $f_s/3$. This deviation from the ideal operation can be attributed to the additional phase shift (or delay) measured in the loop filter. The output spectrum in Fig. 7.10 also shows a significant amount of second, third and fourth order harmonic distortion. As the tuning frequency of the SDM is not equal to $f_s/4$, the third harmonic distortion does not alias into the signal band.

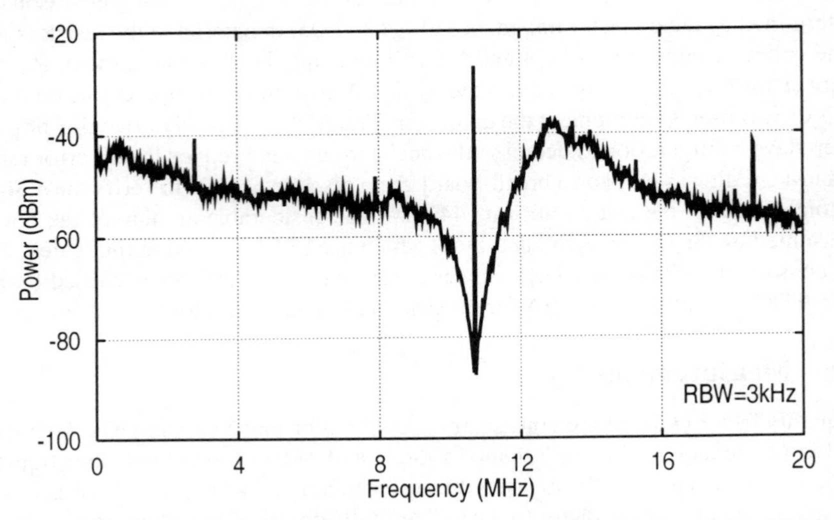

Figure 7.10: Typical output spectrum of the fourth order bandpass SDM.

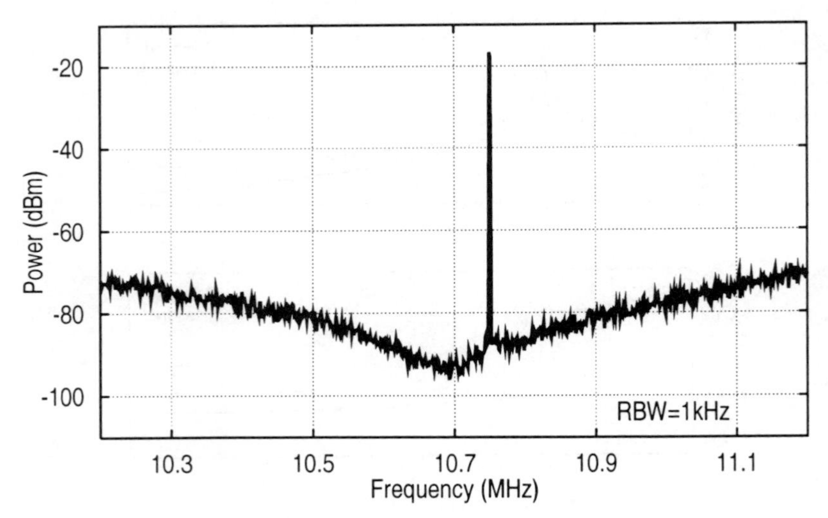

Figure 7.11: In-band region of the output spectrum.

The measured SNDR as a function of the input power is shown in Fig. 7.12 for a signal bandwidth of 9kHz and 200kHz. The maximum SNDR equals 54.4dB for a 200kHz signal bandwidth and 62dB for a 9kHz bandwidth. The slope of the SNDR curve is not unity. The SNDR degrades more rapidly when the input power is decreased. Additionally, the SNDR measured in a bandwidth of 9kHz approaches the SNDR measured in a 200kHz bandwidth when the input power is decreased. This means that for low input power levels, the in-band quantization noise increases and is located near the signal frequency. This behavior may have (at least) two causes. First, the output power spectra were measured using an analog spectrum analyzer connected to an NRZ digital output of the one bit quantizer. As it was mentioned in sec. 4.4.3, limited rise and fall times in NRZ signals cause signal dependent distortion. In order to eliminate these errors, the output spectra should be determined by an FFT of the output samples. At the time of the measurements however, a logic analyzer was not available and the digital output samples of the SDM could not be stored. Secondly, an increase of the quantization noise near the input signal frequency may be caused by jitter in the clock signal. The errors in the sample period causes errors in the apparent frequency of the input signal which can be seen in the output spectrum as phase noise near the input signal frequency.

Although a significant amount of harmonic distortion can be seen in the output spectrum of Fig. 7.10, the third order intermodulation is low. Figure 7.13 shows the output spectrum of the SDM with two tones applied to the input. The IM3 is less than -75dB relative to the carrier power of -20dBm for a carrier distance of $\Delta f = 200$kHz. A summary of the performance of the discrete fourth order bandpass SDM is given in Table 7.2.

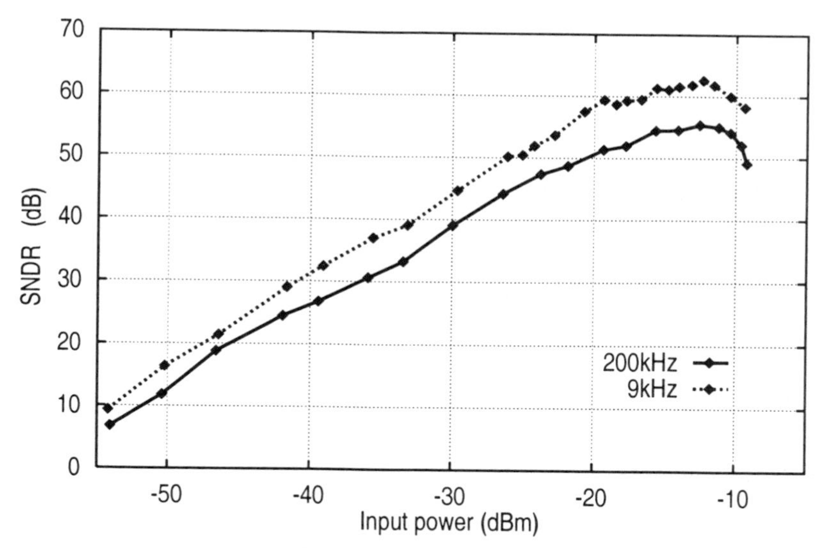

Figure 7.12: Measured SNDR vs. input power characteristic.

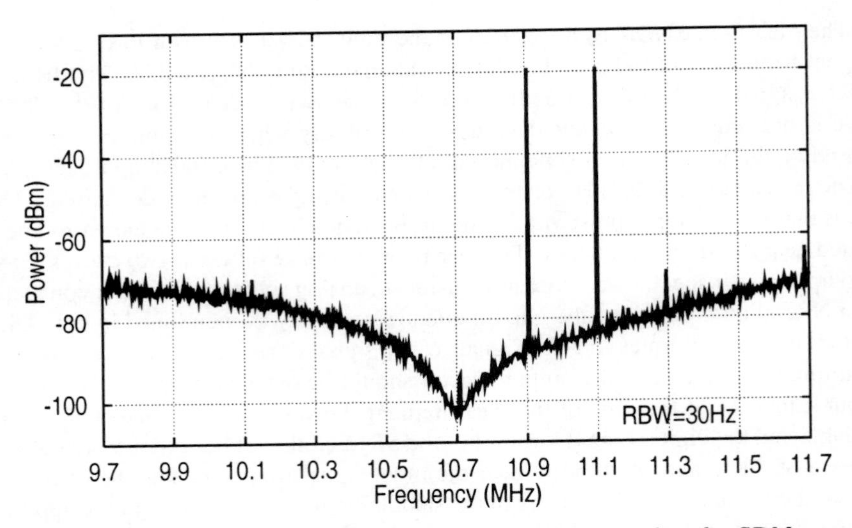

Figure 7.13: *Two tone IM3 measurement of the fourth order SDM.*

Table 7.2: *Discrete 4th Order Bandpass ΣΔ Performance*

supply voltage	5V
power consumption	440mW
sample frequency	40MHz
tuning frequency	10.7MHz
peak SNDR (BW=200kHz)	54.4dB
peak SNDR (BW=9kHz)	62dB
IM3 ($\Delta f = 200$kHz)	-75dBc

7.3 Fully Integrated Sixth Order bandpass SDM [69]

In this section the design and implementation of a fully integrated sixth order CT bandpass SDM is presented. The design method of the SDM is similar to the design of the discrete fourth order SDM of the previous section.

7.3.1 Discrete Time Filter Design

For the discrete time design, the loop filter of the sixth order SDM is chosen from the class of filters described by (5.70). The sixth order DT loop filter can be written as:

$$G(z) = \frac{(1+ae^{j\theta_0}z^{-1})^3(1+ae^{-j\theta_0}z^{-1})^3}{(1+be^{j\theta_0}z^{-1})^3(1+be^{-j\theta_0}z^{-1})^3} - 1 \qquad (7.11)$$

Similar to the fourth order SDM, the tuning frequency is f_0=10.7MHz and the sample frequency is $f_s = 40$MHz. The resulting angular tuning frequency is $\theta_0 = 1.681$ rad.

The poles will be placed on the unit circle: $b = 1$. The position of the zeroes is determined by the stability of the modulator. According to the rule of thumb (see Fig. 5.28) the parameter a should be at least $a \geq 0.68$ for small signal stability. In order for the modulator to be stable for a considerable input signal amplitude, the value of a should be larger than this minimum value. In this case, a value of $a = 0.75$ is chosen. The resulting root locus for $\alpha = 0$ is shown in Fig. 7.14. The large signal stability is de-

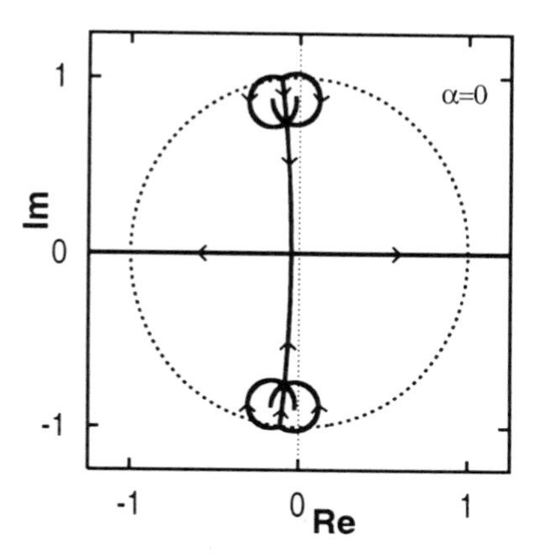

Figure 7.14: *Root locus of sixth order bandpass SDM ($\alpha = 0$).*

termined by the two unit circle crossings of the root locus near the tuning frequency of $\theta_0 = 1.681$ rad. The worst-case value of the quantizer gain equals $\lambda = 0.2231$, corresponding to a maximum quantizer input amplitude of:

$$A_{x,\max} \approx \frac{\frac{\pi}{4} \cdot \frac{q}{2}}{\lambda} = 2.85q \tag{7.12}$$

Simulations show that the maximum quantizer input amplitude which does not cause instability equals $A_{x,\max} = 2.64q$. This maximum quantizer input amplitude results in a maximum SDM (in-band) sine-wave input amplitude of $A_{i,\max} = 0.33q$ or -6.6dB relative to the quantizer output power $q^2/4$. The error in the prediction of the maximum quantizer input amplitude can be explained by the fact that the phase uncertainty was not taken into account (see sec. 5.7). Taking into account this phase uncertainty results in a predicted value of $A_{x,\max} = 2.56q$.

The noise transfer function $\mathrm{NTF}(z) = 1/(1 + G(z))$ of the SDM with loop filter (7.11) is shown in Fig. 7.15 for the designed values $a = 0.75$, $b = 1$, and $\theta = 1.681$. The noise suppression at the tuning frequency is much larger than in the case of the fourth order bandpass SDM (see Fig. 7.3). Again the out-of-band gain of the NTF exceeds two (or 6dB), violating LEE's stability rule. Simulations and the implementation reported below show that this bandpass SDM is stable for small input signals.

When using the linear prediction method described in sec. 4.1, the maximum achievable SNR can be determined. In a bandwidth of 200kHz, the 6th order SDM achieves a maximum SNR of 93dB. Simulations show that the maximum SNR indeed is 93dB in 200kHz. A typical (simulated) output spectrum is shown in Fig. 7.16.

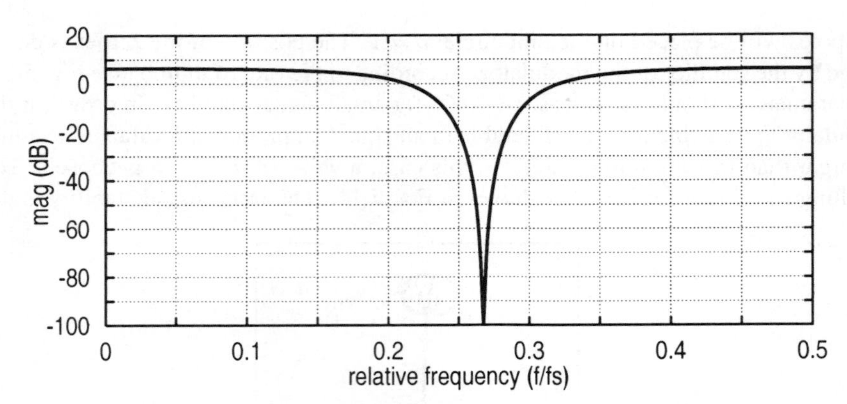

Figure 7.15: NTF of the sixth order bandpass SDM.

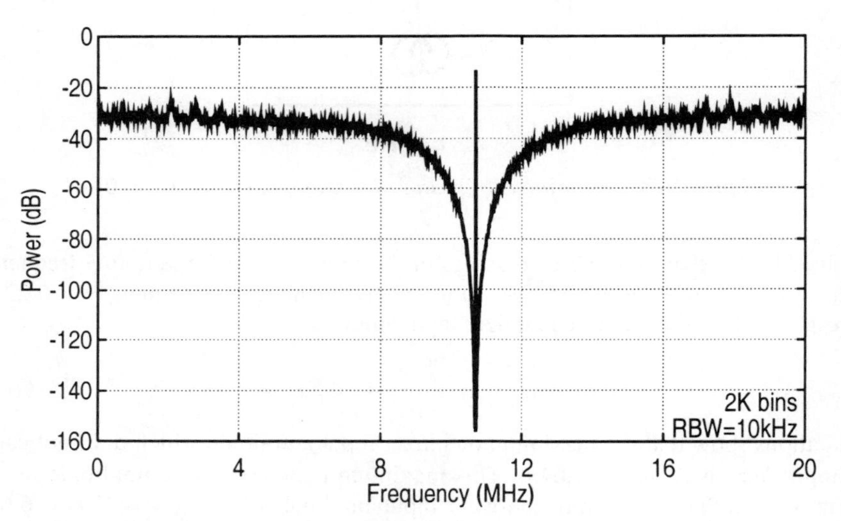

Figure 7.16: Typical output spectrum (simulation) of a sixth order SDM with loop filter (7.11) with $a = 0.75$, $b = 1$ and $\theta_0 = 1.681$. Input power is -13dB relative to the quantizer output power.

7.3.2 Continuous Time Filter Design

Similar to the fourth order SDM in the previous section, the loop filter structure from Fig. 6.8 is used. The resulting SDM structure is shown in Fig. 7.17. The six poles of the loop filter are realized by three resonators. The zeroes of the loop filter are realized by placing the resonators in cascade with six feed forward paths. The coefficients of the feed forward paths are realized by resistors R_1 to R_6. The coupling resistors R_{c1} and R_{c2} are used for scaling the voltages within the filter to the same level. In the case that the quality factor of the resonators is high (i.e. $Q > 100$), the transfer function of

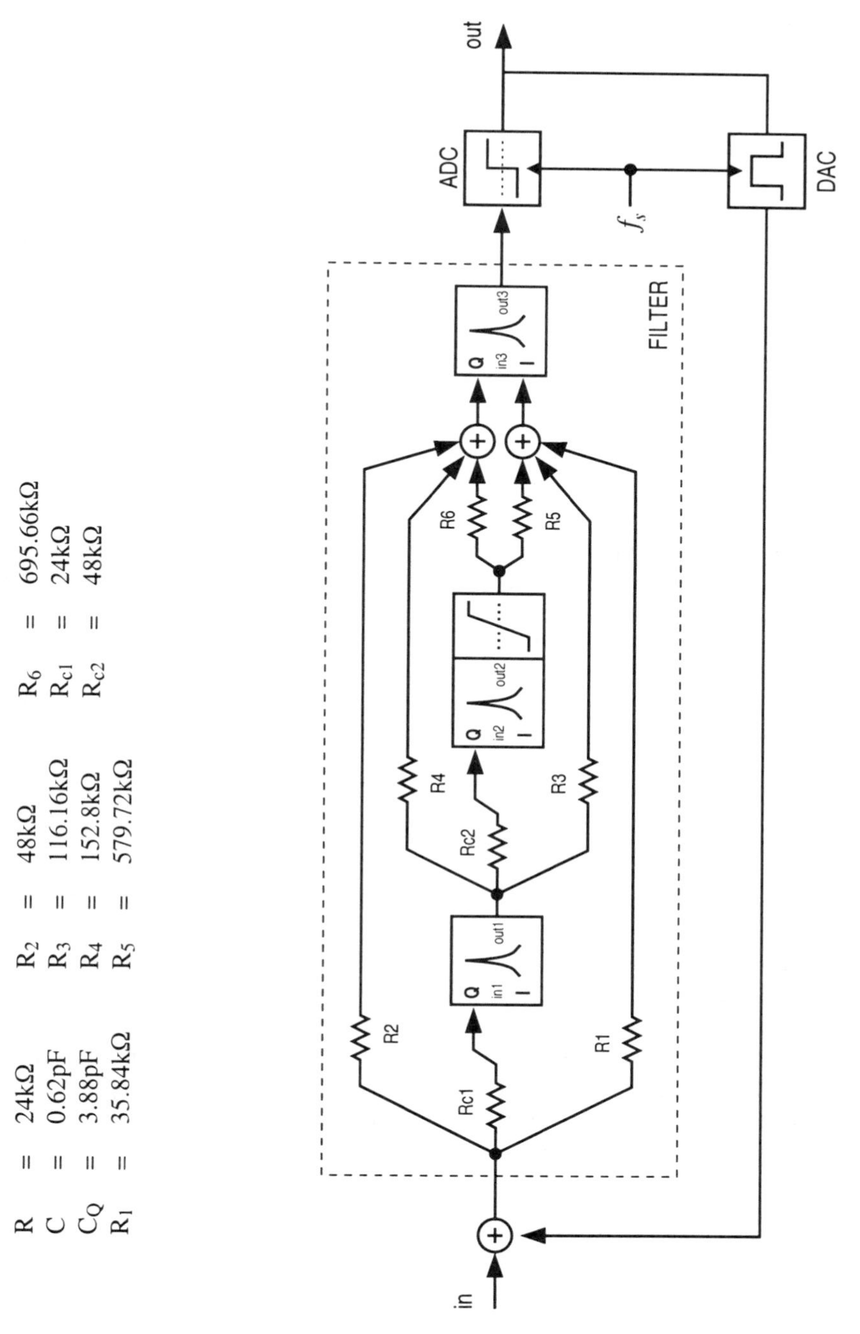

Figure 7.17: *Diagram of the sixth order bandpass SDM.*

the loop filter of the SDM can be written as:

$$F(p) = \omega_0 \frac{N(p)}{(p^2 + p\omega_0/Q + \omega_0^2)^3} \tag{7.13}$$

with:

$$N(p) = c_2 p^5 + \omega_0(c_1 + c_4)p^4 + \omega_0^2(2c_2 + c_3 + c_6)p^3 + \omega_0^3(2c_1 + c_4 + c_5)p^2 + \omega_0^4(c_2 + c_3)p + \omega_0^5 c_1 \tag{7.14}$$

The coefficients $c_1 \ldots c_6$ are defined by:

$$\begin{aligned} c_1 &= \tfrac{R}{R_1} & c_2 &= \tfrac{R}{R_2} \\ c_3 &= \tfrac{R}{R_{c1}} \cdot \tfrac{R}{R_3} & c_4 &= \tfrac{R}{R_{c1}} \cdot \tfrac{R}{R_4} \\ c_5 &= \tfrac{R}{R_{c1}} \cdot \tfrac{R}{R_{c2}} \cdot \tfrac{R}{R_5} & c_6 &= \tfrac{R}{R_{c1}} \cdot \tfrac{R}{R_{c2}} \cdot \tfrac{R}{R_6} \end{aligned} \tag{7.15}$$

in which R is the resistance used in the resonator (see subsection 7.3.3; implementation). The coefficients can be solved by calculating the equivalent DT transfer function $G_{eq}(z)$ and matching it with the desired DT transfer function $G(z)$ of eq. (7.11). As in the case of the fourth order bandpass SDM of the previous section, an RTZ pulse with duration of half the sample period T_s is chosen for the DAC pulse shape (see Fig, 7.7). Substituting $\omega_0 = 2\pi f_0$ with f_0 =10.7MHz, the CT to DT transformation $F(p) \rightarrow G_{eq}(z)$ can be solved using Maple [68]. Matching $G_{eq}(z)$ with $G(z)$ then gives:

$$\begin{aligned} c_1 &= -1.339 & c_2 &= 1.000 \\ c_3 &= -0.413 & c_4 &= 0.314 \\ c_5 &= -0.041 & c_6 &= 0.035 \end{aligned} \tag{7.16}$$

Stability at large input signal amplitudes is ensured when the output of the second resonator is limited, as it is indicated in Fig. 7.17. When the limiter is active, the contribution of the second resonator to the overall transfer of the loop filter is reduced to a DC bias and the effective order of the filter is reduced to four. Stable operation is ensured when the resulting fourth order modulator is also stable. As the feedback loop of the SDM is not entirely broken by the operation of the limiter, the performance of the SDM will degrade gracefully [59]. The stability of the resulting fourth order modulator can be checked by setting $c_5 = 0$ and $c_6 = 0$ and calculate the resulting equivalent DT loop filter transfer function:

$$G_{eq,4}(z) = \frac{d_1 z^{-1} + d_2 z^{-2} + d_3 z^{-3} + d_4 z^{-4}}{(1 + 2\cos(\theta_0) z^{-1} + z^{-2})^2} \tag{7.17}$$

with

$$\begin{aligned} d_1 &= -0.2249 \\ d_2 &= -1.7866 \\ d_3 &= -0.3965 \\ d_4 &= -1.0583 \end{aligned} \tag{7.18}$$

Checking the small signal stability with the rule of thumb of section 5.6.4 shows that the resulting fourth order modulator indeed is stable. At an angular frequency $\theta = \pi/3$ and $\theta = 2\pi/3$ the radii of the roots in the root locus are always inside the unit circle, as can be seen from the root locus with $\alpha = -1$ shown in Fig. 7.18.

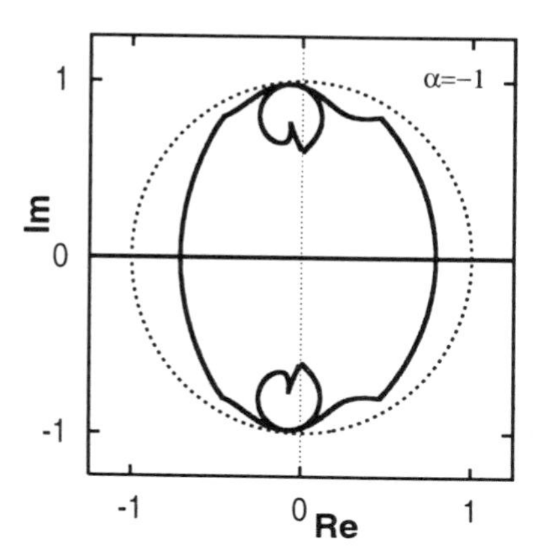

Figure 7.18: Root locus ($\alpha = -1$) of the fourth order SDM resulting from clipping the second resonator in the sixth order SDM.

7.3.3 Implementation

The resonators in the loop filter of the sixth order bandpass SDM are implemented using the balanced integrator shown in Fig. 7.19. A balanced integrator (BI) resonator is preferred over a transconductance capacitor (gmC) resonator as it has a larger linear output range. The capacitors are placed in a feedback around the operational amplifier so that parasitic capacitances to the substrate have little influence. The coefficients are implemented using passive resistors. In the resonator of Fig. 7.19, resistors R and capacitors C determine the tuning frequency of the resonator. The capacitors C_Q limit the quality factor of the resonator; the additional (dashed) transistors allow tuning of the resonator. The balanced implementation allows realization of negative coefficients by reversing positive and negative terminals. The ADC and DAC of the quantizer are

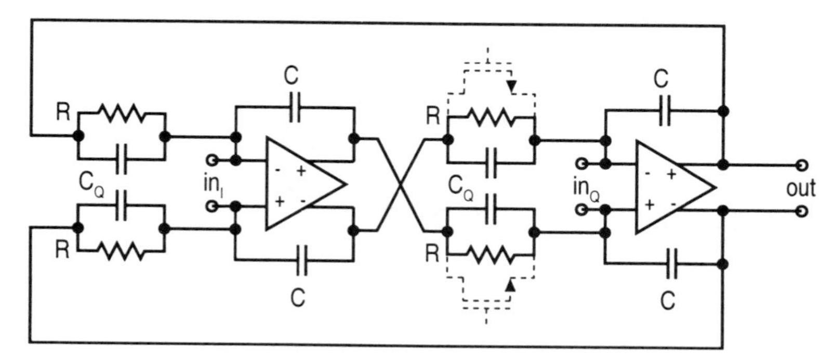

Figure 7.19: Balanced integrator resonator. Tuning transistors are dashed.

also implemented fully differentially, improving the noise and distortion immunity. The supply voltage is 3.3V for the digital part and 5V for the analog sections. The common mode DC level of the differential signals is 3V and the amplitude is $3V_{pp}$ maximally.

The transconductance amplifier of the resonator is shown in Fig. 7.20 (see [70]). The cross-coupled transistors M_1 and M_2 increase the transconductance of the differential pair formed by transistors M_3 and M_4. The amplifier has a gain-bandwidth product of 1.1GHz. The parasitic delay caused by the poles of the transconductance amplifier increases the quality factor of the resonators from the designed value of $Q = 100$ to $Q = 180$. The transconductance amplifiers and resistors cause noise in the filter. The equivalent input noise power density of the filter at the tuning frequency equals $8.4 \cdot 10^{-8}$ V/$\sqrt{\text{Hz}}$. Combined with a quantizer output power of −3.5dBm[2] (see below), the noise of the filter results in a maximum achievable SNR of 72dB in 200kHz. Due to the high Q of the resonators, the noise caused by the filter will exceed the theoretical quantization noise and serve as a dither signal for the SDM.

The quantizer ADC (see Fig. 7.21) consists of two flip-flops in a master slave configuration that reduces the bit error rate without increasing the parasitic load on the filter. Two non-overlapping clock signals are generated internally by dividing the external clock.

The DAC of the quantizer is shown in Fig. 7.22. It consists of a logic block to create RTZ signals, a cascoded differential stage and two source followers. Two dummy

[2]In this book, 0 dBm refers to the *voltage* that gives 1mW across 50Ω. Actual impedance levels may vary.

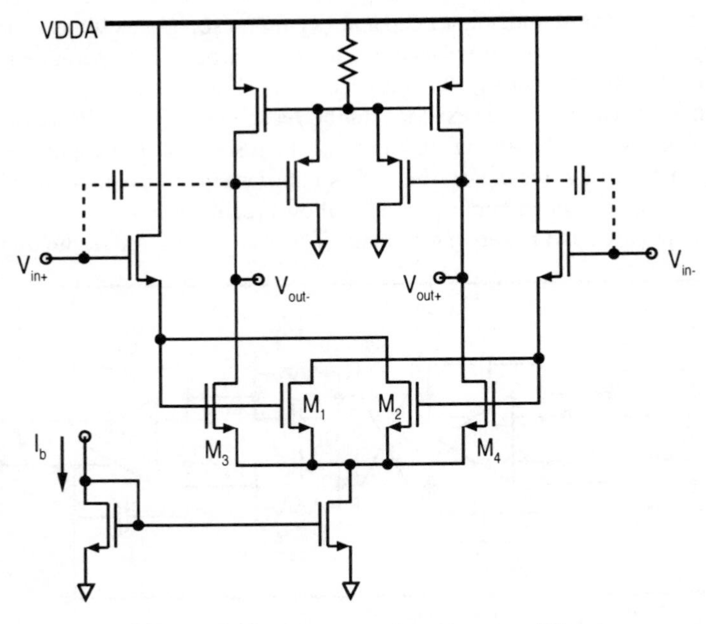

Figure 7.20: *Transconductance amplifier.*

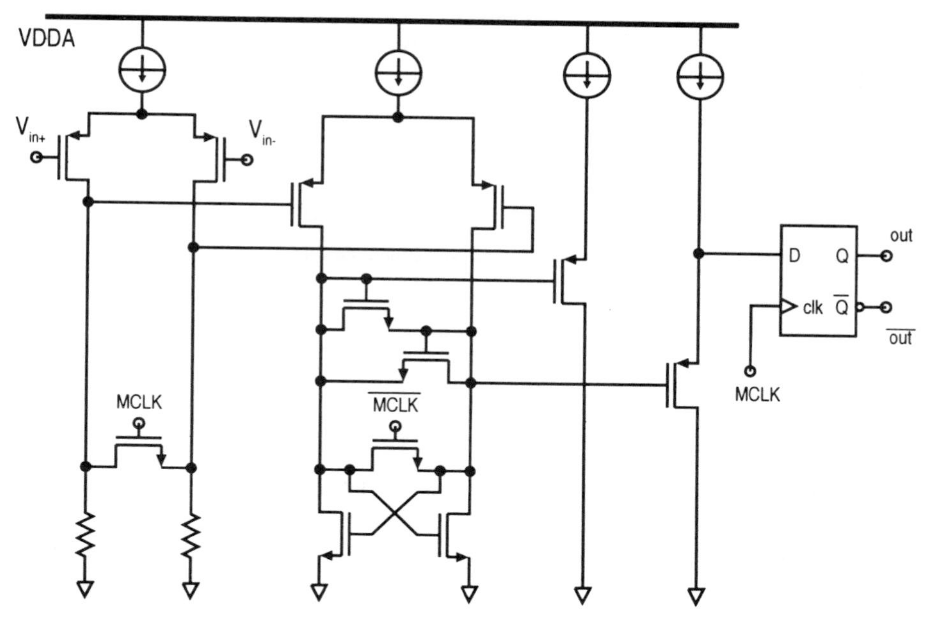

Figure 7.21: One bit ADC.

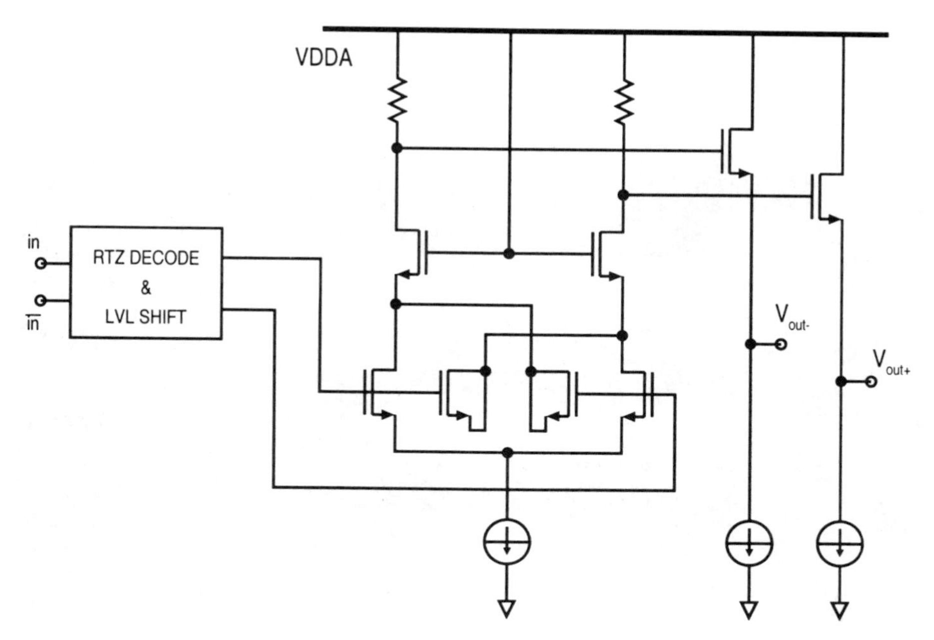

Figure 7.22: One bit DAC.

transistors are added to the differential stage to reduce glitches caused by charge storage. The RTZ pulses of the quantizer DAC have a width of $T_s/4$ and an amplitude of $1.2V_{pp}$ which gives -3.5dBm (see Fig. 7.23). As the CT filter was designed us-

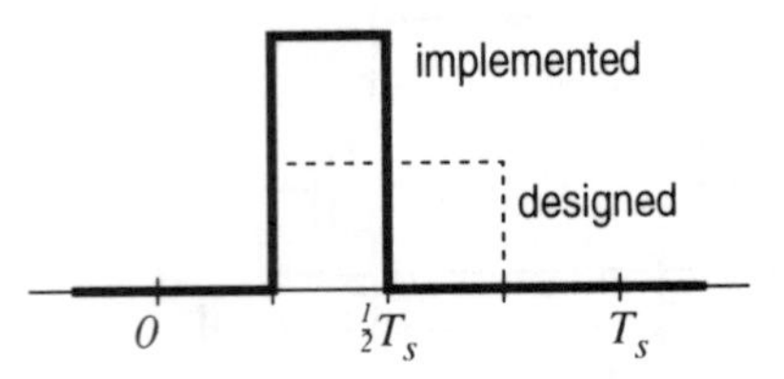

Figure 7.23: Designed and actually implemented DAC pulse.

ing RTZ pulses of $T_s/2$, a narrower DAC pulse changes the equivalent DT transfer function slightly. The loss of power of the DAC pulse is compensated by doubling the amplitude of the DAC pulse and so avoids a reduction of the performance. As the modification to the RTZ pulse slightly reduces the delay of the signal in the loop, the stability of the modulator will not be affected detrimentally.

In order to monitor the individual outputs of the resonators in the loop filter, three analog buffers are also included in the design of the SDM. Two versions of the SDM were made: a tunable version with the tuning transistors in the resonators and a non tunable version without the tuning transistors. Both versions of the SDM are realized in 0.5μm double-poly CMOS. A die photograph is shown in Fig. 7.24. The corresponding floor plan is shown in Fig. 7.25. The loop of the SDM is closed externally. The core circuit (analog filter, ADC and DAC) measures 0.9x0.4mm^2 and consumes 60mW at a sample rate of 40MHz. The digital buffers consume 9.5mW at 40MHz

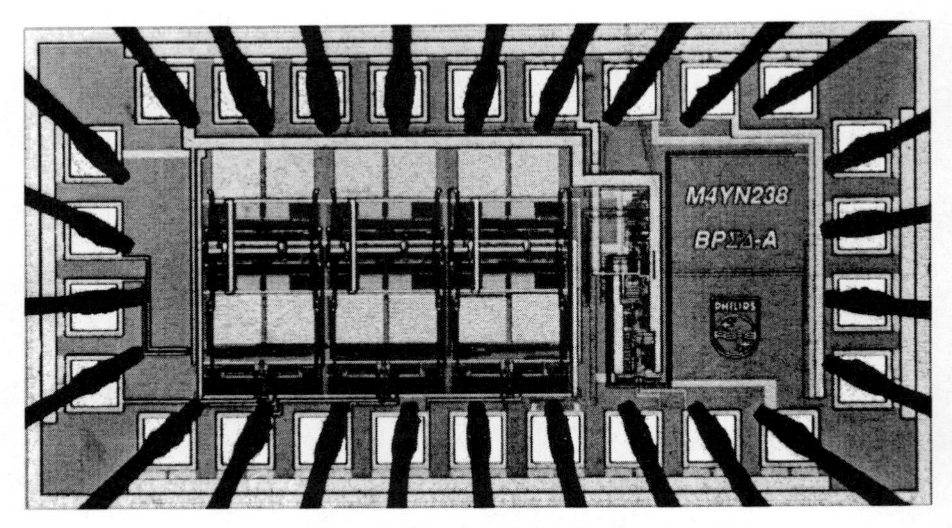

Figure 7.24: Die photo of the sixth order SDM IC.

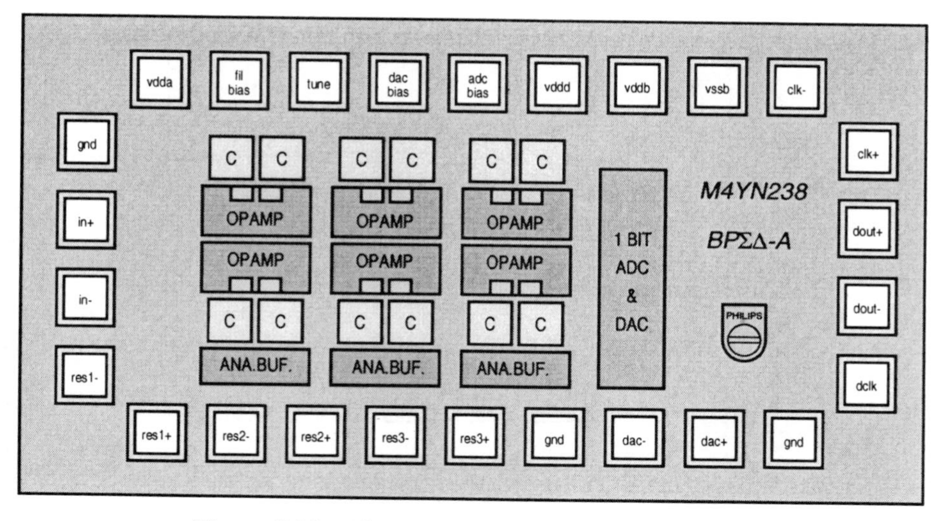

Figure 7.25: Floor plan of the sixth order SDM IC.

with a load capacitance of 8pF. The analog buffers consume 117mW. The total power consumed by the chip equals 186.5mW at 40MHz. As most of the power is consumed by the analog filter and the output buffers, the sample frequency is insignificant for the total power consumption.

7.3.4 Measurements

The first check on the 6th order bandpass SDM IC is the verification of the stability of the modulator. Both the tunable and non tunable version of the SDM proved to be stable under all measurement operating conditions. Input signal amplitudes well exceeding the DAC output levels causes a severe degradation in the performance of the SDM, but the limiter at the output of the second resonator ensures that the modulator is free of large signal limit cycles. The sample frequency can be varied from 30MHz to 80MHz without resulting in large signal limit cycles.

Non tunable version

The non tunable version of the sixth order bandpass SDM was typically tuned at 9.15MHz. The large deviation from the designed value of 10.7MHz is caused by the fact that the tuning frequency depends on absolute resistance and capacitance values. The production process may cause up to 30% variation in absolute parameter values. A tuning mechanism in continuous time bandpass modulators is therefore necessary. Note that the filter coefficients depend on the ratio between two or more resistance values. As a relative accuracy of 1% can be obtained for resistance and capacitance ratios, tuning of the feedforward resistors in Fig. 7.17 is not required. The measured transfer characteristic of a single resonator is shown in Fig. 7.26. The measured quality factor of the resonator is $Q = 180$. The overall filter characteristic is shown in Fig. 7.27.

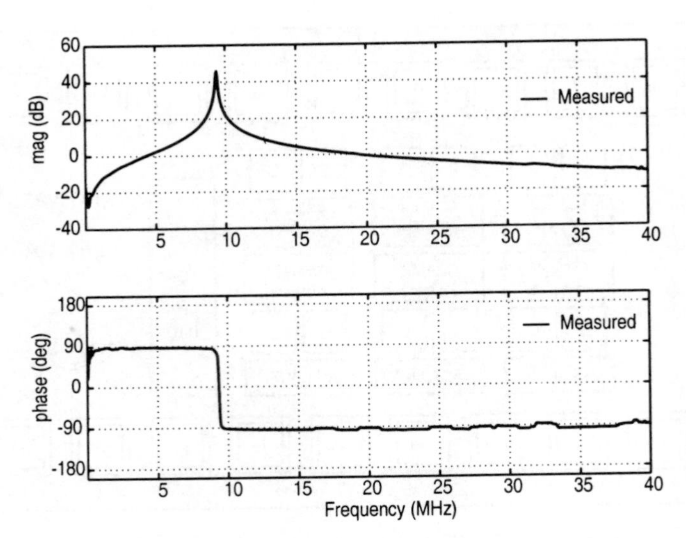

Figure 7.26: *Measured resonator transfer characteristic of the non tunable version.*

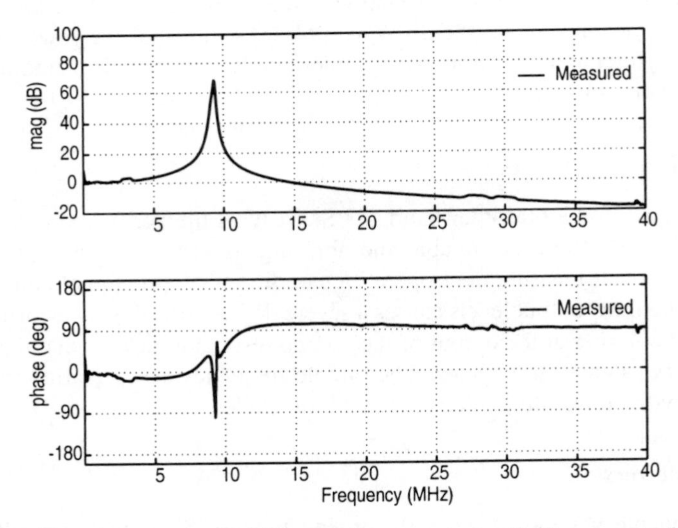

Figure 7.27: *Measured filter transfer characteristic of the non tunable version.*

The performance of the SDM was determined by calculating the FFT of the output data. Up to 512k samples of digital output samples of the modulator were stored in a logic analyzer and transferred to a computer for further analysis. The non tunable version has an input referred idle channel noise of -78.5dBm in 200kHz and -92.5dBm in 9kHz, resulting in a dynamic range of 72dB and 86dB respectively. A typical output spectrum of the non tunable SDM is shown in Fig. 7.28. The SNDR vs. input power characteristic is shown in Fig. 7.29. At an input of -6dB relative to the DAC output power, the non tunable version achieves a maximum SNDR of 67dB in 200kHz and 80dB in 9kHz. The corresponding ENOB are 10.8 bit and 13 bit. The crossings of the characteristics with the SNDR=0dB line give an estimate of the dynamic range (DR). The curve corresponding to a measurement bandwidth of 9kHz is less smooth than the curve corresponding to 200kHz. This is caused by the resolution of the measurement. Using a 512k FFT for signals sampled at 40MHz gives an FFT bin bandwidth of approximately 76Hz. For the SNDR measurement in a bandwidth of 9kHz only 118 bins are used. The SNDR measurement in a bandwidth of 200kHz uses over 2000 bins,

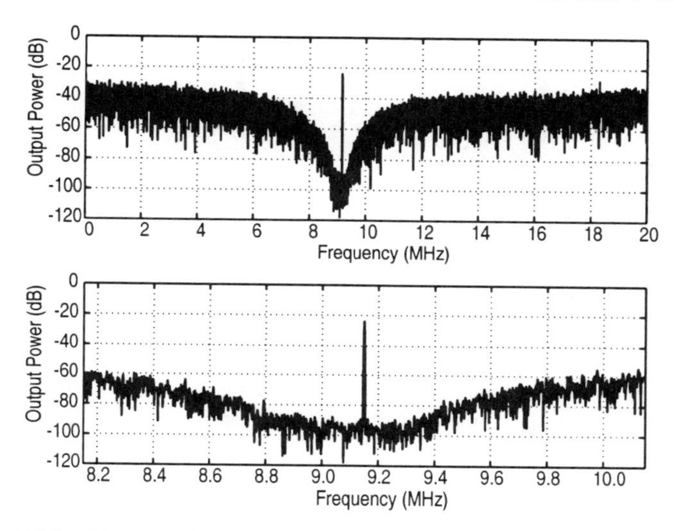

Figure 7.28: Measured output spectrum of non tunable version (32k FFT).

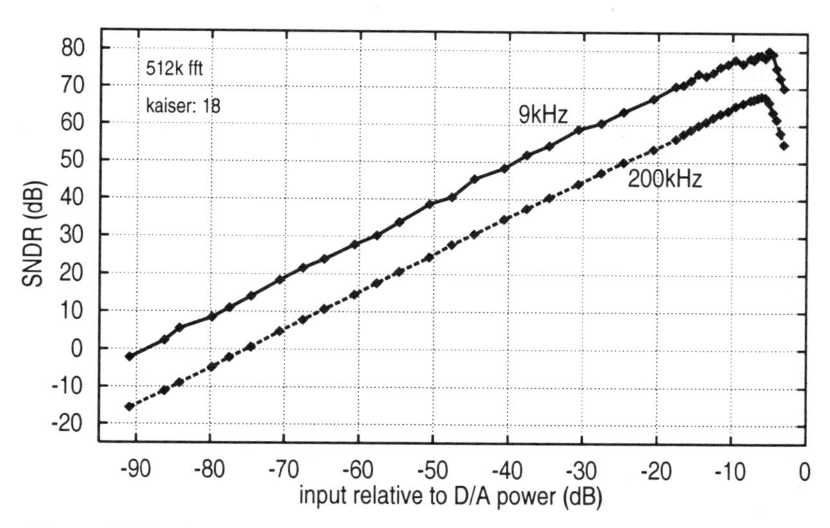

Figure 7.29: SNDR vs. input power characteristic of the non tunable SDM.

thus giving a more accurate result. A total of 26 samples of the non tunable SDM IC were available for testing. This number allows some statistical measurements. In Figure 7.30 the histograms are shown for the maximum SNDR measured in 200kHz and the absolute tuning frequency. The samples having the lowest maximum SNDR

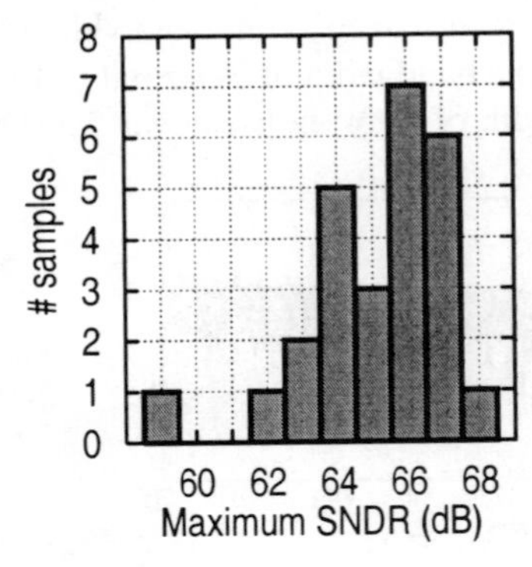

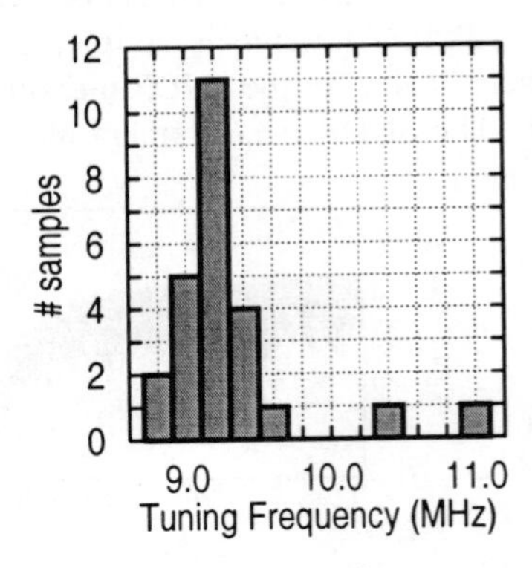

Figure 7.30: *Histograms of the maximum SNDR and tuning frequency of 26 samples of the non tunable SDM IC.*

correspond to the two samples with tuning frequencies 10.4MHz and 11.0MHz. These samples clearly deviate from the expected normal distribution and may be affected by manufacturing process defects.

In order to determine the third order intermodulation distortion (IM3), a two-tone measurement is performed. Two input signals ('carriers') with frequencies $f_0 - \Delta f$ and $f_0 - 2\Delta f$ (or $f_0 + \Delta f$ and $f_0 + 2\Delta f$) are applied to the modulator, and the spurious response at the tuning frequency f_0 is measured as a function of the carrier amplitude level. In Fig. 7.31 the resulting characteristic is shown for three different carrier spacings.

The optimal IM3 of -82dBc is reached at -13dB carrier power, corresponding to an IP3 of +24.5dBm. For carrier levels lower than -16dB, the IM3 intermodulation product is near the measurement noise floor. Note that the measurement noise floor is not horizontal, as the vertical scale is relative to the carrier level (in dBc). When the carrier level decreases, the distance between the noise floor and the carrier also becomes smaller. Although the absolute noise floor remains constant, the relative noise floor increases as the carrier power is decreased. For carrier powers exceeding -11dB, the IM3 intermodulation distortion increases rapidly. At such high input powers, the signals inside the loop filter become large. The limiter of the second resonator in the loop filter will be active, increasing the (intermodulation) distortion. An output spectrum of the modulator during the two-tone test is shown in Fig. 7.32. The carrier spacing is 200kHz.

The performance of the non tunable modulator has also been measured for sample frequencies up to 80MHz. An increase of the sample frequency did not result in an increase of the performance. This indicates that the SNDR of the non tunable version is limited by the noise of the loop filter.

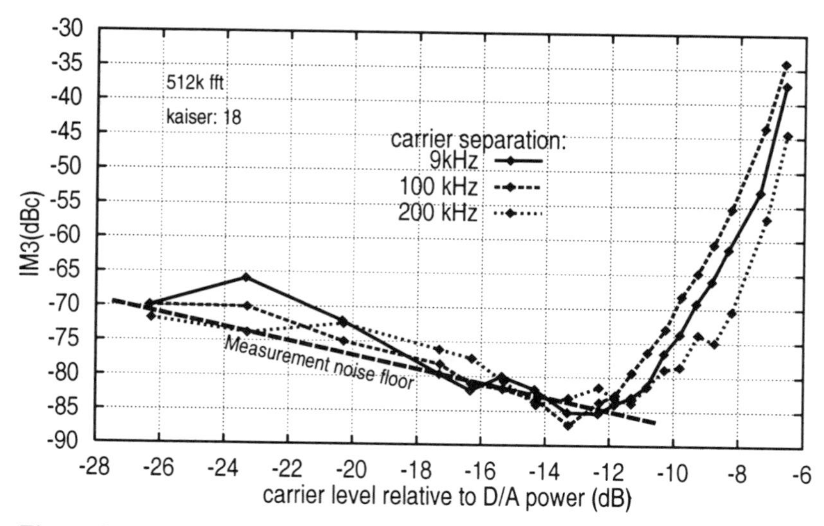

Figure 7.31: IM3 vs. carrier power characteristic of the non tunable SDM for three different carrier spacings.

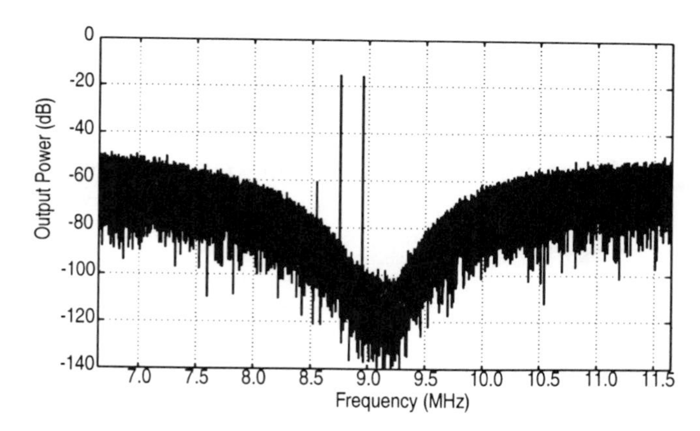

Figure 7.32: Output spectrum of the two-tone IM3 measurement of the non tunable SDM (512k FFT; Δf =200kHz).

Tunable version

In the tunable version of the sixth order bandpass SDM, an additional transistor is used to allow tuning of the loop filter. The transistors are placed in parallel to the resistors inside the balanced integrator resonators (see Fig. 7.19). By changing the gate-source voltage of the transistor, the conductance of the drain source path can be changed. The source of the transistor is connected to an input node of the transconductance amplifier. Due to the feedback, this node is a virtual 'ground'. Consequently, the conductance of the tuning transistor can be changed by changing the gate voltage. The tuning frequency of the loop filter was set to 10.7MHz.

The measured transfer characteristics of a single resonator and the total filter are shown in Fig. 7.33 and Fig. 7.34 respectively. The designed resonator and filter characteristics are also shown in the same figures. The quality factor of the resonators of the tunable version is equal to the quality factor of the non tunable version: $Q = 180$. The characteristic of the resonator shows little deviation of the designed magnitude and phase response. The measured loop filter characteristic shows a considerable loss

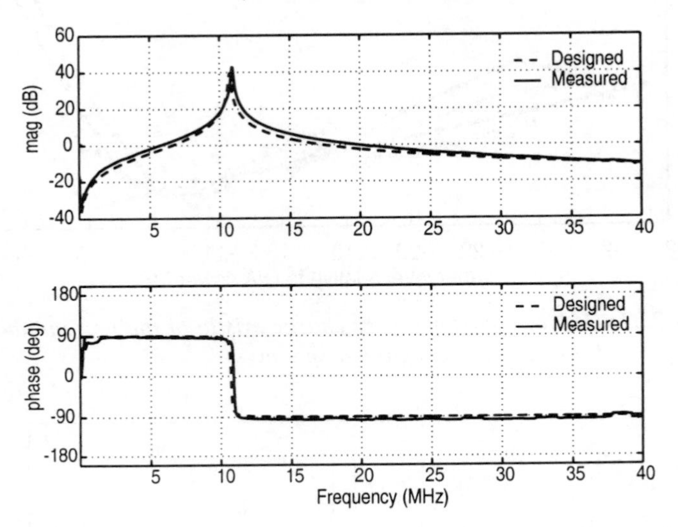

Figure 7.33: *Measured resonator transfer characteristic of the tunable version.*

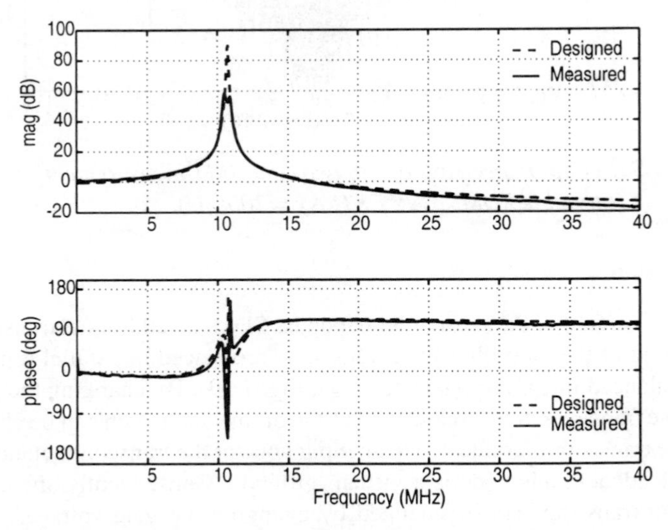

Figure 7.34: *Measured filter transfer characteristic of the tunable version.*

of gain with respect to the designed value. This is mainly due to the fact that the three resonators do not have identical tuning frequencies. The magnitude response of the loop filter clearly shows two bumps in the pass band of the filter. The phase response of the loop filter agrees with the designed characteristic. This is important as the phase characteristic of the loop filter mainly determines the stability of the modulator. This is due to the fact that the stability of the modulator is determined by the zeroes of the loop filter. As the zeroes are not located near the unit circle, they do not have a significant effect on the amplitude response of the filter.

The tunable version of the sixth order SDM has an input referred idle channel noise of -73.5dBm in 200kHz and -87.5dBm in 9kHz. This is significantly (5dB) higher than the idle channel noise of the non tunable version. The resulting dynamic range is also lower: 67dB in 200kHz and 81dB in 9kHz. A typical output spectrum of the tunable version is shown in Fig. 7.35. The spectrum of the tunable version not only shows a

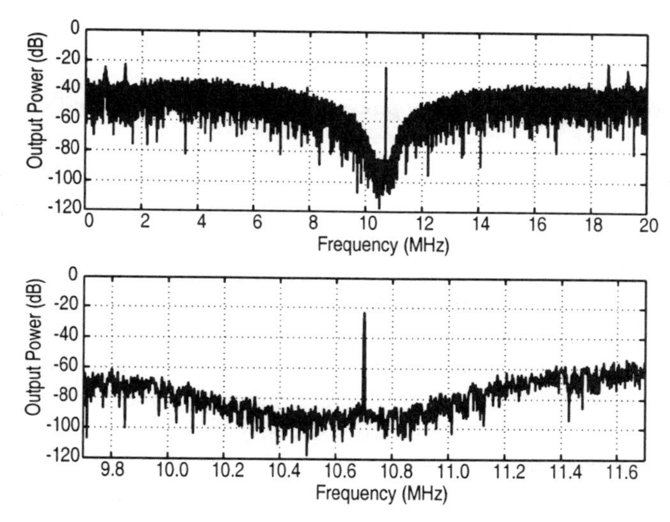

Figure 7.35: *Measured output spectrum of tunable version (32k FFT).*

higher noise floor than the output spectrum of the non tunable version, but also shows larger distortion components near 1MHz and 19MHz. The nonlinearity of the tuning transistors in the resonators clearly introduces additional distortion which affects the performance of the SDM. In Fig. 7.36 the SNDR vs. input power characteristic of the tunable version is shown. For an input of -6dB relative to the DAC output power, the maximum SNDR equals 63.5dB in 200kHz and 76dB in 9kHz. The corresponding ENOB are 10.2 bit for 200kHz and 12.3 bit for 9kHz. As in the case of the non tunable version, the SNDR curve corresponding to a measurement bandwidth of 9kHz is less smooth than the curve corresponding to a measurement bandwidth of 200kHz. For an input power between -30dB and -20dB, the 9kHz SNDR curve shows a significant drop in the SNDR. It is unclear whether this drop is caused by measurement limitations or is caused by the nonlinearity of the tuning transistors. In order to investigate the reason for this drop, more output data of the SDM should be used for the calculation of the

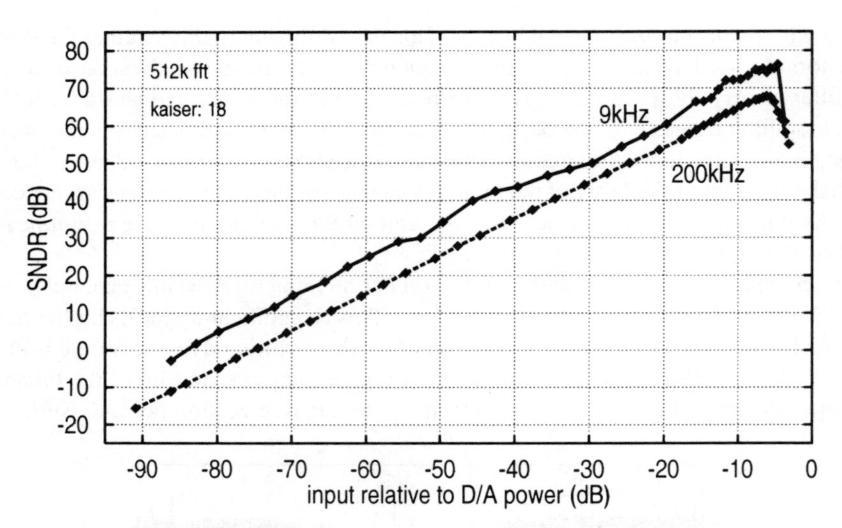

Figure 7.36: *SNDR vs. input power characteristic of the tunable SDM.*

SNDR. A real-time digital filter could be used to remove the out-of-band quantization errors. In that case, the SNDR can be calculated from the decimated output signal. Unfortunately, such a digital filter was not available at the time of the measurements.

For the tunable version 27 samples were available for testing. In Fig. 7.37 a histogram of the maximum SNDR is shown. Apart from one sample with a very poor performance, most of samples achieve a similar maximum SNDR.

The third order intermodulation (IM3) of the tunable version is determined using a two-tone measurement. The resulting IM3 as a function of the carrier input levels is shown in Fig. 7.38 for three different carrier spacings. A typical output spectrum of the two-tone IM3 measurement is shown in Fig. 7.39. The optimal IM3 is -75dBc at an input carrier level of -14dB. The nonlinearity of the tuning transistors gives a reduction

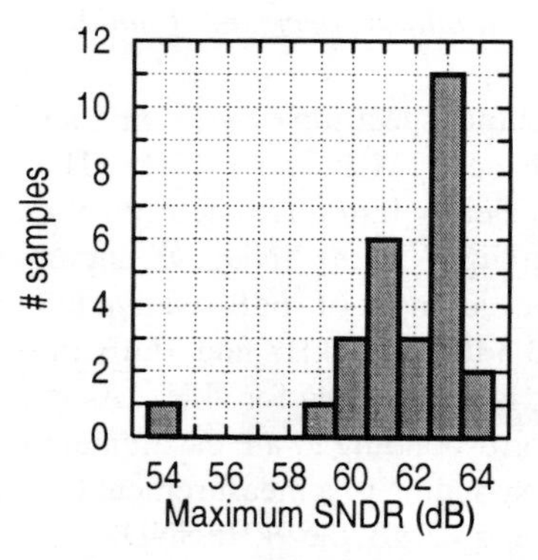

Figure 7.37: *Histogram of the maximum SNDR of 27 samples of the tunable sixth order bandpass SDM IC.*

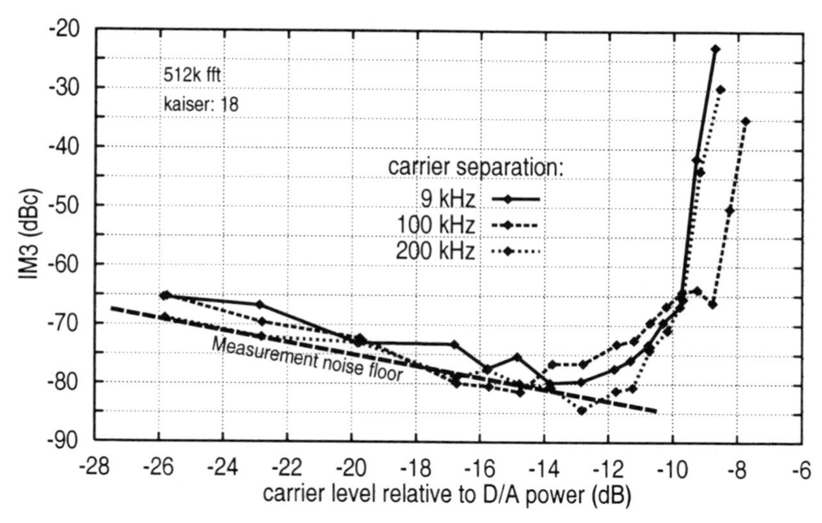

Figure 7.38: *IM3 vs. carrier power characteristic of the tunable SDM for three different carrier spacings.*

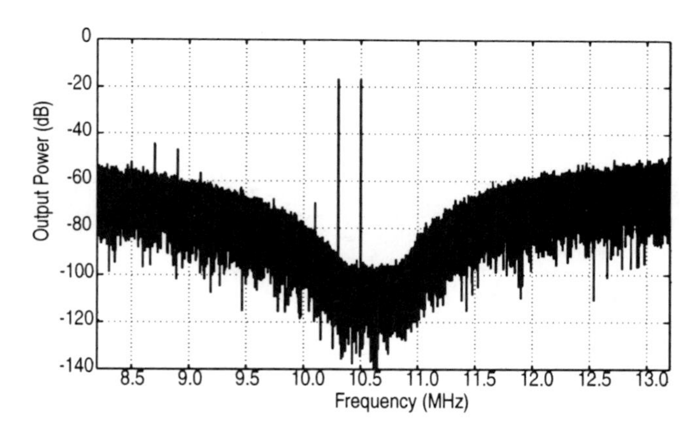

Figure 7.39: *Output spectrum of the two-tone IM3 measurement of the tunable SDM (512k FFT; Δf =200kHz).*

of only 6dB in the IM3 compared to the non tunable version. The nonlinearity causes distortion in the output of the loop filter. As it was explained in sec. 4.4.3, the distortion that adds to the output of the loop filter is suppressed by the feedback loop. For the tunable version, the out-of-band intermodulation distortion components are much larger than for the non tunable version (see Fig. 7.32). For reason of the suppression by the feedback loop, the *in-band* IM3 distortion components for the tunable version are only slightly larger than in the case of the non-tunable version.

The performance of non tunable modulator has also been measured for sample frequencies up to 80MHz. An increase in the sample frequency did not result in an increase of the performance. This indicates that the SNDR of the tunable version is limited by the noise of the loop filter and the distortion caused by the non linearity of the tuning transistors.

Summary

Table 7.3 shows a summary of the performance of the sixth order bandpass SDM. The power consumption does not include the consumption by the analog output buffers which are not required for the operation of the SDM.

Table 7.3: *6th Order Bandpass SDM Performance*

	non tunable		*tunable*	
technology	0.5μm CMOS			
supply voltage	5.0V analog, 3.3V digital			
power consumption	60 mW @ f_s=40MHz			
sample frequency	30 - 80 MHz			
tuning frequency	9.15 MHz		10.7 MHz	
	Bandwidth		Bandwidth	
	200kHz	9kHz	200kHz	9kHz
idle channel noise	-78.5dBm	-92.5dBm	-73.5dBm	-87.5dBm
DR	72dB	86dB	67dB	81dB
peak SNDR	67dB	80dB	63.5dB	76dB
ENOB	10.8	13	10.2	12.3
IM3 rel. to carriers	-82dBc		-75dBc	
max. diff. input	200mV			

7.3.5 Further Remarks

The design of the sixth order bandpass SDM shows that stable high order bandpass modulators can be designed and can achieve good performance. The design of the SDM was not optimized for power, performance or linearity. For example, the amplifiers in the loop filter are identical. The power consumption of an amplifier is mostly but not exclusively determined by the gain-bandwidth product and the noise requirements. Although the gain-bandwidth requirements of all the amplifiers are similar, the noise requirements are not. The noise contribution of the total loop filter is mainly given by the noise of the first resonator. As the performance of the SDM seems to be limited by the noise of the loop filter, the performance can be increased by reducing the noise of the first resonator. Normally, this results in an increase in the power consumption of these amplifiers. The total power consumption of the loop filter does not have to increase as the power consumption of the amplifiers in the second and third resonator may be decreased. Another approach to increase the SNDR is the replacement of the first resonator by an (external) LC resonator.

As it has been shown in Table 7.3, the tuning transistors have a considerable impact on the overall performance of the SDM. When comparing the performance of the tunable and non tunable version, the difference of the SNDR and IM3 are clear. The tuning mechanism of the loop filter can be improved in several ways. An example of a tuning mechanism reducing the effects of the nonlinear tuning transistors is by discrete switching of the value of the resistor or capacitor inside the resonator (see Fig. 7.40).

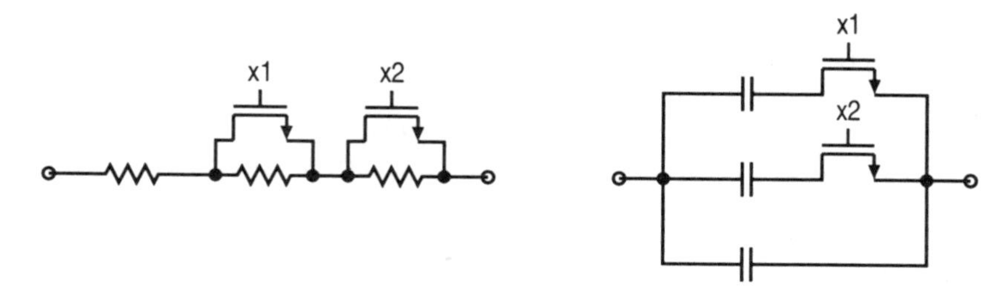

Figure 7.40: *Tuning by discrete switching of resistor and capacitor values.*

As the transistors are now used as switches, the effect of the transconductance nonlinearity of the transistor is reduced significantly. Other methods to reduce the effects of nonlinearity, such as adding compensation transistors are also applicable.

The modulator can also be improved by other techniques such as implicit input filtering (see sec. 3.6.3). The signal transfer function of the SDM can be designed to suppress out-of-band signals, thereby improving the intermodulation distortion. The Signal Transfer Function of the sixth order bandpass SDM equals:

$$\mathrm{STF}(z) = \frac{G(z)}{1+G(z)} \tag{7.19}$$

with the loop filter $G(z)$ according to (7.11) with $a = 0.75$, $b = 1$ and $\theta_0 = 1.681$. The STF is shown in Fig. 7.41. The STF is entirely determined by the design of the NTF. By making a separate feedforward path in the loop filter for the SDM input signal and the DAC feedback signal, the zeroes of the STF can be placed independently from the

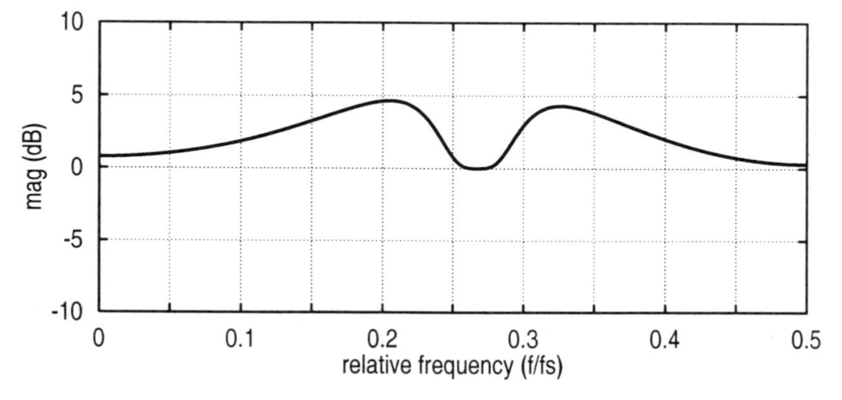

Figure 7.41: *STF of the sixth order bandpass SDM.*

NTF. (see sec. 3.6.3). The STF is changed into:

$$\mathrm{STF}(z) = \frac{H(z)}{1+G(z)} \tag{7.20}$$

The poles of the transfer function $H(z)$ are the poles of $G(z)$. The zeroes of $H(z)$ can be chosen freely. In the case of a sixth order SDM, $H(z)$ has five zeroes. By placing a zero of $H(z)$ on the unit circle, the signal transfer is suppressed for the corresponding frequency. As an example the zeroes of $H(z)$ are placed at $z = 0$, $z = e^{\pm j\theta_1}$ and $z = e^{\pm j\theta_2}$. The angular frequencies $\theta_{1,2}$ are chosen symmetrical around the angular tuning frequency θ_0. Here, $\theta_{1,2} = 2\pi f_{1,2}/f_s$ with $f_1 = (10.7-4)$MHz and $f_2 = (10.7+4)$MHz, giving $\theta_1 = 1.052$ and $\theta_2 = 2.309$. The resulting STF with implicit input filtering is shown in Fig. 7.42. The signal suppression by the zeroes on the unit circle is clearly visible. Other out-of-band signals are suppressed by more than 25dB. The exact location of the zeroes of $H(z)$ allows a trade-off between the out-of-band signal suppression and the width of the center lobe of the STF. The main lobe can be made smaller by placing the zeroes closer to the tuning frequency at the cost of a lower out-of-band signal suppression.

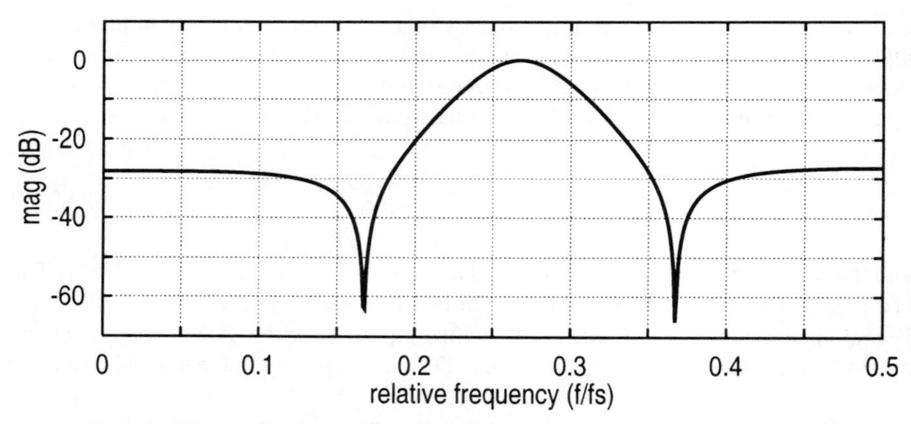

Figure 7.42: *STF of the 6th order bandpass SDM with implicit input filtering.*

7.4 Comparison

In this section, the fourth order bandpass SDM of section 7.2 and the fully integrated sixth order bandpass SDM of section 7.3 are compared to several published bandpass sigma delta modulators. The overall performance of the SDMs is evaluated by a figure of merit (FOM) [71]. Such a figure should include at least the achieved SNDR or DR, the signal bandwidth BW and the power consumption P:

$$\mathrm{FOM} = \frac{4k_bT \cdot \mathrm{SNDR} \cdot \mathrm{BW}}{P} \tag{7.21}$$

in which k_b is Boltzmann's constant ($k_b = 1.38 \cdot 10^{-23}$ J/K) and T is the temperature in Kelvin (K). Note that the SNDR should not be given in dB when using eq. (7.21). In the case of bandpass modulators, the absolute tuning frequency also affects the overall merit of the SDM. In table 7.4 the specifications of a number of 'recently' published bandpass SDMs are given.

The examination of the figures of merit for the listed implementations shows that the FOM value varies by more than three orders of magnitude. This is caused by the fact that a wide range of implementations is selected. Not only are the implementations realized in several process technologies, but the implementations are also intended for different applications. Some applications require a large bandwidth with a low SNDR; other applications require a large SNDR and a small bandwidth. For the SDM, the trade-off between SNDR and bandwidth is not proportional. An SDM intended for a large bandwidth and low SNDR application will have a lower FOM. For example: A second order lowpass SDM achieves a theoretical SNDR of 80dB at OSR=100 and an SNDR of 55dB at OSR=20. For a sample frequency of 1MHz and a power consumption of 1mW, the figure of merit equals FOM=$8.3 \cdot 10^{-6}$ and FOM=$1.3 \cdot 10^{-7}$ respectively. Clearly, a different choice of oversampling ratio leads to a variation in the FOM of nearly two orders of magnitude. One could suggest to include the effect of a higher order loop filter into the FOM. In that case however, the FOM is no longer compatible with other types of ADCs such as flash converters.

Several types of bandpass modulators have been developed. The most common

Table 7.4: *Comparison of published bandpass sigma delta modulators.*

Ref.	Type	order	OSR	f_s (MHz)	f_0 (MHz)	BW (kHz)	SNDR (dB)	Power (mW)	FOM
[67]	LC*	4	63	10	2.5	80	53	-	-
[72]	LC†	4	65	26	6.5	200	55	1350	$7.8 \cdot 10^{-10}$
[73]	SC	4	99	1.852	0.455	8	63	240	$1.1 \cdot 10^{-9}$
[58]	SC	4	120	7.2	1.8	30	74	-	-
[74]	SC	2	107	42.8	10.7	200	55	60	$1.7 \cdot 10^{-8}$
[75]	SC/TI	4	133	8	2	30	56	1.6	$1.2 \cdot 10^{-7}$
[76]	T/H	2	125	250	65.2	1000	40	350	$4.7 \cdot 10^{-10}$
[77]	gmC	4	32	160	5	2500	63	-	-
[78]	SC/TI	4+2‡	32	13	3.25	200	64	14.4	$5.8 \cdot 10^{-7}$
[79]	SC/TI	4	200	80	20	200	75	72	$1.5 \cdot 10^{-6}$
[80]	BI	2	2000	4000	55.5	1000	80	1400	$1.2 \cdot 10^{-6}$
[60]	gmC	2	500	200	50	200	46	-	-
[81]	C-SC	4	25	10	3.75	200	62	130	$4.0 \cdot 10^{-8}$
[82]	LC	2	9500	3800	950	200	57	135	$1.2 \cdot 10^{-8}$
[83]	DCV	4	1000	400	100	200	54	330	$2.5 \cdot 10^{-9}$
[84]	DCV	2	250	20	400	40	70	18	$3.7 \cdot 10^{-7}$
sec. 7.2	LCgmC*	4	100	40	10.7	200	54	440	$1.9 \cdot 10^{-9}$
sec. 7.3	BI	6	100	40	10.7	200	67	60	$2.8 \cdot 10^{-7}$

* Discrete Implementation. † External LC resonator. ‡ two stage cascaded SDM.

LC	LC resonator	**SC**	Switched Capacitor
SC/TI	Switched Capacitor / Time Interleaved	**T/H**	Track and Hold
gmC	Transconductance Capacitor CT filter	**BI**	Balanced Integrator CT filter
C-SC	Complex Switched Capacitor	**DCV**	Direct Conversion

type of loop filter is the switched capacitor (SC) discrete time loop filter. For most of these SDMs, the tuning frequency is equal to a quarter of the sample frequency: $f_0 = f_s/4$. The implementations using an SC filter achieve a relatively low FOM. A modified version of the SC filter uses a time-interleaving (SC/TI) technique. By using two lowpass SC loop filters which are sampled using time-interleaved clocks, an overall bandpass characteristic can be obtained. The implementations that use this technique (SC/TI) achieve a relatively high FOM. This is due to the fact that the time-interleaved sampling causes the SC filters to process baseband signals. The lowpass SC filters can achieve better settling and noise performance than their true bandpass counterparts. Another type of SDM, the direct-conversion (DCV) SDM [83], uses a mixer inside the loop for down-conversion of the signal frequencies to cope with the high accuracy requirements at high signal frequencies. The implementations of such a DCV SDM also achieve a relatively high FOM. Although the SC loop filter is the dominantly used filter type for bandpass SDMs, the implementations using continuous-time loop filters such as Balanced Integrator (BI) or transconductance-capacitor (gmC) filters are capable of achieving a comparable FOM. The implementation described in [82] shows that an SDM with a fully integrated LC resonator achieves an acceptable SNDR at tuning frequencies near 1GHz.

Together with the SDM described in [85], the SDM of sec. 7.3 of this book shows that the signal-dependent stability of high order, one bit SDMs can be dealt with and that these SDMs are a viable solution for high-performance low-power A/D conversion of IF signals

7.5 Conclusions

In this chapter, three implementations of sigma delta modulators were presented. The digital hardware set-up was used to speed up the simulation of the SDM by an order of magnitude compared to software simulations. Lengthy simulations are needed to verify the signal dependent stability of high order modulators.

The discrete fourth order bandpass SDM described in sec. 7.2 shows the feasibility of continuous-time bandpass SDMs with a mixed active/passive loop filter. The fourth order SDM was used as a test case for the design of the fully integrated sixth order bandpass SDM described in sec. 7.3. This sixth order SDM is one of the first single loop bandpass SDM with a loop filter order higher than four. A test IC shows that the signal-dependent stability of such a high order bandpass SDM does not have to be a barrier in the quest for higher performance and lower power consumption of SDMs. The measured results of this sixth order bandpass SDM are compared to the performance of reported bandpass SDMs. This comparison shows that SDMs using a CT filter can achieve a similar performance at approximately the same power consumption. A CT implementation of the loop filter has the advantage that the SDM is easily tunable and that the tuning frequency does not depend on the sample frequency.

CHAPTER 8

CONCLUSION AND DISCUSSION

The study described in this book has been focused on the performance, stability and design aspects of sigma delta modulators (SDMs). The study shows once more that the apparent simplicity of error feedback coders such as noise shapers and sigma delta modulators is deceitful. The presence of the nonlinearity of the quantizer and a loop filter in the feedback loop is problematic for an exact analysis. In this book it has been shown that a considerable amount of properties can be derived by the use of linearization techniques.

The linear modeling of the quantization errors provides an accurate prediction for the signal to noise ratio (SNR) of sigma delta modulators [29, 31]. Deviations in the actual SNR performance are explained by an analysis of the deterministic behavior of the modulator. In-band tones are responsible for bumps and slope changes in the SNR vs. input power characteristics of the SDM (section 4.2).

The stability of an SDM has been analyzed using the describing function method. The modeling of the quantizer by a single global gain was shown to be inadequate for the prediction of idle patterns and zero-input stability of an SDM (section 5.3). Careful analysis revealed that a sampled quantizer exhibits a phase uncertainty (section 5.4). The incorporation of (an approximation of) the phase uncertainty in the describing function model improved the prediction of idle patterns. The extended model predicted the small-signal stability boundaries for loop filter parameters of lowpass modulators within 5% of the experimental values. Similar results are obtained with a rule of thumb. The rule is based on the presented theoretical analysis and so avoids the hasty generalization of a restricted set of empirical results.

As to the stability of bandpass SDMs, the accuracy of our rule of thumb varies. For a fourth order bandpass SDM, our rule of thumb does not give an accurate prediction of the small-signal stability boundary. For bandpass SDMs of order higher than four, the rule of thumb provides an accurate (within 5%) estimate of the small signal stability boundary for 60% of the tuning range of the SDM (between DC and $f_s/2$). (section 5.6). Overall, the proposed stability analysis technique provides an intuitive and accurate approach to the stability problems of SDMs. The phase uncertainty of a sampled quantizer is shown to have a considerable impact on the stability of sigma delta modulators.

The design of high order (> 4) continuous time bandpass sigma delta modulators is strongly affected by the stability properties of the SDM. The continuous time loop filter should have sufficient degrees of freedom for the implementation of the desired (small-signal stable) discrete time SDM (section 6.4). A set of bandpass filter structures that are based on a cascade of resonators has been proposed (section 6.6). Sigma delta modulators employing continuous time loop filters can be subsampled, i.e. the tuning frequency of the loop filter is higher than half the sample frequency. A high ratio between the tuning frequency and the sample frequency increases the requirements on the quality factor of the loop filter. Therefore, the use of subsampling appears to be restricted to moderate subsampling ratios (section 6.5).

Using the described stability analysis technique, three SDM implementations were realized. First, an all-digital programmable hardware set-up was used for SDM simulations. The hardware set-up reduces the simulation time by an order of magnitude and allows real-time evaluation of SDMs with time-varying input signals (section 7.1). Second, a discrete implementation of a fourth order bandpass SDM shows the feasibility of bandpass SDMs with a mixed active/passive loop filter. Third, a fully integrated sixth order bandpass SDM was designed. It is based on the given stability criteria and the proposed filter structures (section 7.3). This sixth order SDM is one of the first implementation of a single loop SDM with a loop filter order higher than four. Together with the SDM described in [85], the measurement results of a 0.5μm CMOS test IC show that (section 7.4):

- the signal dependent stability of such a SDM does not pose a barrier in the quest of high performance and low power bandpass ADCs;
- intermediate frequency (IF) bandpass SDMs using continuous time loop filters can achieve a similar performance as their switched capacitor counterparts at approximately the same power consumption.

Discussion

As in the case of nearly all research studies, certain aspects of the research topic remain unexplored and many questions remain unanswered. The rule of thumb for small signal stability boundaries was derived for a class of lowpass modulators. Even though direct application of the rule to bandpass SDMs provides an accurate small-signal stability prediction in the majority of cases, the assumptions used for the derivation of the rule of thumb are violated. Further investigations of the stability properties of bandpass SDMs tuned to frequencies near $f_s/3$ and $f_s/6$ that might lead to a more refined rule of thumb is recommended. Also, the relationship between the input signal (amplitude) and the gain λ and phase uncertainty α of the quantizer model is not discussed in this book. Experiments show that the quantizer input signal increases with an increasing input signal amplitude. This corresponds to a decreasing value for λ. However, an analytical expression for the relationship between the model parameters and the input signal as was derived in [31] is not available using this model. Finally, the noise model described by STIKVOORT [29], the noise/stability model introduced by ARDALAN and PAULOS [31] and the describing function stability model described in this book share common features such as singularities in the solutions. The phase uncertainty in the describing function model introduces a new aspect to the relationship between the three models. The investigation of this relationship may lead to new insights for the stability of bandpass modulators.

REFERENCES

[1] J.C. Candy and G.C. Temes, "Oversampling methods for A/D and D/A conversion," in *Oversampling Delta-Sigma Data Converters*, J.C. Candy and G.C. Temes, Eds. IEEE Press, New York, 1991.

[2] Steven R. Norsworthy, Richard Schreier, and Gabor C. Temes, Eds., *Delta-Sigma Data Converters; Theory, Design and Simulation*, IEEE press, New York, 1997, ISBN 0-7803-1045-4.

[3] W.R. Bennett, "Spectra of quantized signals," *Bell System Technical Journal*, vol. 27, pp. 446–472, 1948.

[4] Bernard Widrow, "A study of rough amplitude quantization by means of nyquist sampling theory," *IRE Trans. on Circuit Theory*, pp. 266–276, Dec. 1956.

[5] Anekal B. Sripad and Donald L. Snyder, "A necessary and sufficient condition for quantization errors to be uniform and white," *IEEE Trans. on Acoustics, Speech, and Signal Processing*, vol. 25, no. 5, pp. 442–448, Oct. 1977.

[6] Stanley P. Lipshitz, Robert A. Wannamaker, and John Vanderkooy, "Quantization and dither: A theoretical survey," *J. Audio Eng. Soc.*, vol. 40, no. 5, pp. 355–375, May 1992.

[7] Nelson M. Blachman, "The intermodulation and distortion due to quantization of sinusoids," *IEEE Trans. on Acoustics, Speech and Signal Processing*, vol. 33, no. 6, pp. 1417–1426, Dec. 1985.

[8] T.A.C.M. Claasen and A. Jongepier, "Model for the power spectral density of quantization noise," *IEEE Trans. on Acoustics, Speech and Signal Processing*, vol. 29, no. 4, pp. 914–917, Aug. 1981.

[9] A.W.M. van den Enden and N.A.M. Verhoekx, *Discrete-time signal processing*, Prentice Hall, 1989.

[10] Rudy van de Plassche, *Integrated Analog-to-Digital and Digital-to-Analog Converters*, Kluwer Academic Publishers, Dordrecht, NL, 1994, ISBN 0-7923-9436-4.

[11] E.M. Deloraine, S. van Miero, and B. Derjavitch, "Méthode et système de transmission par impulsions," Brevet d'invention 932.140, August 10 1948, (in French).

[12] F. de Jager, "Delta modulation - a method of PCM transmission using the one unit code," *Philips Res. Rep.*, vol. 7, pp. 442–466, 1952.

[13] C.C. Cutler, "Transmission systems employing quantization," U.S. Patent No. 2,927,962, March 8 1960 (filed in 1954).

[14] H.A. Spang and P.M. Schultheiss, "Reduction of quantization noise by use of feedback," *IRE Trans. on Commun. Syst.*, pp. 373–380, Dec. 1962.

[15] H. Freeman, *Discrete-time Systems*, John Wiley & sons, inc., New York, 1965.

[16] R. Schreier and M. Snelgrove, "Bandpass sigma-delta modulation," *Electronic Letters*, vol. 25, pp. 1560–1561, Nov. 1989.

[17] H. Inose, Y. Yasuda, and J. Murakami, "A telemetering system by code modulation - $\Delta - \Sigma$ modulation," *IRE Trans. Space Electron. Telemetry*, vol. SET-8, pp. 204–209, Sept. 1962.

[18] Toshio Hayashi, Yasunobu Inabe, Kuniharu Uchimura, and Tadakatsu Kimura, "A multistage delta-sigma modulator without double integration loop," in *IEEE Int. Solid-State Circuits Conf. (ISSCC) digest of Technical Papers*, Feb. 1986, pp. 182–183.

[19] Yasuyuki Matsuya, Kunihari Uchimura, Atsushi Iwata, Tsutomu Kobayashi, Masayuki Ishikawa, and Takeshi Yoshitome, "A 16-bit oversampling A-to-D conversion technology using triple integration noise shaping," *IEEE Journal of Solid-State Circuits*, vol. 22, no. 6, pp. 921–929, Dec. 1987.

[20] P. Aziz, H. Sorensen, and J. van der Spiegel, "Multiband sigma-delta modulation," *Electronic Letters*, pp. 760–762, Apr. 1993.

[21] Ian Galton and Hendrik T. Jensen, "Oversampling parallel delta-sigma modulator A/D conversion," *IEEE Trans. on Circuits and Systems-II*, vol. 43, no. 12, pp. 801–810, Dec. 1996.

[22] Ramin Khoini-Poorfard, Lysander B. Lim, and David A. Johns, "Time-interleaved oversampling A/D converters: Theory and practice," *IEEE Trans. on Circuits and Systems-II*, vol. 44, no. 8, pp. 634–645, Dec. 1997.

[23] Dagnachew Birru, "Reduced sample rate sigma delta modulation using recursive deconvolution," *Int. Journal of Circuit Theory and Applications*, vol. 25, pp. 419–437, 1997.

[24] L. Risbo, "Improved stability and performance from sigma-delta modulators using one-bit vector quantization," in *Proc. Int. Symp. Circuits and Systems (ISCAS)*, Chicago, USA, May 1993, pp. 1365–1368.

[25] S. Jantzi, K. Martin, M. Snelgrove, and A. Sedra, "A complex bandpass $\Delta\Sigma$ converter for digital radio," in *Proc. Int. Symp. Circuits and Systems (ISCAS)*, June 1994, pp. 453–456.

[26] James C. Candy, "A use of double integration in sigma delta modulation," *IEEE Trans. on Communications*, vol. 33, no. 3, pp. 249–258, Mar. 1985.

[27] Richard Schreier and Bo Zhang, "Delta-sigma modulators employing continuous-time circuitry," *IEEE Trans. on Circuits and Systems I: Fundamental Theory and Applications*, vol. 43, no. 4, pp. 324–332, Apr. 1996.

[28] Bhagawati P. Agrawal and Kishan Shenoi, "Design methodology for ΣΔM," *IEEE Trans. on Communications*, vol. 31, no. 3, pp. 360–370, Mar. 1983.

[29] Eduard F. Stikvoort, "Some remarks on the stability and performance of the noise shaper or sigma-delta modulator," *IEEE Trans. on Communications*, vol. 36, no. 10, pp. 1157–1162, Oct. 1988.

[30] Eduard Stikvoort, *Some subjects in Digital Audio, Noise Shaping, Sample-Rate Conversion, Dynamic Range Compression and Testing*, Ph.D. thesis, Eindhoven University of Technology, Sept. 1992, ISBN 90-9005162-7.

[31] S.H. Ardalan and J.J. Paulos, "An analysis of nonlinear behavior in delta-sigma modulators," *IEEE Trans. on Circuits and Systems*, vol. 34, no. 6, pp. 593–603, June 1987.

[32] Augusto Marques, Vincenzo Peluso, Michel S. Steyaert, and Willy M. Sansen, "Optimal parameters for ΔΣ modulator topologies," *IEEE Trans. on Circuits and Systems-II*, vol. 45, no. 9, pp. 1232–1241, Sept. 1998.

[33] Robert M. Gray, "Spectral analysis of quantization noise in a single-loop sigma-delta modulator with dc input," *IEEE Trans. on Communications*, vol. 37, no. 6, pp. 588–599, June 1989.

[34] Robert M. Gray, Wu Chou, and Ping W. Wong, "Quantization noise in single-loop sigma-delta modulation with sinusoidal inputs," *IEEE Trans. on Communications*, vol. 37, no. 9, pp. 956–967, Sept. 1989.

[35] Vladimir Friedman, "The structure of the limit cycles in sigma delta modulation," *IEEE Trans. on Communications*, vol. 36, no. 8, pp. 972–979, Aug. 1988.

[36] James C. Candy and Oconnell J. Benjamin, "The structure of quantization noise from sigma-delta modulation," *IEEE Trans. on Communications*, vol. 29, pp. 1316–1323, Sept. 1981.

[37] Sundeep Rangan and Bosco Leung, "Quantization noise spectrum of double-loop sigma-delta converter with sinusoidal input," *IEEE Trans. on Circuits and Systems II*, vol. 41, no. 2, pp. 168–173, Feb. 1994.

[38] Ramin Khoini-Poorfard and David A. Johns, "Analysis of ΣΔ modulators with zero mean stochastic inputs," *IEEE Trans. on Circuits and Systems II*, vol. 42, no. 3, pp. 164–175, Mar. 1995.

[39] Richard Schreier, "On the use of chaos to reduce idle-channel tones in delta-sigma modulators," *IEEE Trans. on Circuits and Systems-I*, vol. 41, no. 8, pp. 539–547, Aug. 1994.

[40] Robert C. Hilborn, *Chaos and Nonlinear Dynamics*, Oxford University Press, Inc., New York / Oxford, 1994, ISBN 0-19-58816-6.

[41] Lars Risbo, $\Sigma - \Delta$ *Modulators - Stability Analysis and Optimization*, Ph.D. thesis, Technical University of Denmark, June 1994.

[42] O. Feely and L.O. Chua, "The effect of integrator leak in $\Sigma - \Delta$ modulation," *IEEE Trans. on Circuits and Systems*, vol. 38, no. 11, pp. 1293–1305, Nov. 1991.

[43] Richard Feynman, Robert B. Leighton, and Matthew L. Sands, *The Feynman Lectures on Physics: Commemorative Issue*, pp. 24–2, Addison-Wesley Publishing Company, 1989, ISBN 0-201-51003-0.

[44] D.P. Atherton, *Stability of Non-Linear Systems*, Research Studies Press; Wiley, 1981, ISBN 0-471-27856-4.

[45] Saber N. Elaydi, *An Introduction to difference equations*, Springer, Berlin, 1996, ISBN 0-387-94582-2.

[46] A. Lyapunov, "Problème général de la stabilité du movement," in *Ann. of Math Study #17*, Princeton, 1947.

[47] S. Hein and A. Zakhor, "On the stability of sigma delta modulators," *IEEE Trans. on Signal Processing*, vol. 41, pp. 2322–2348, July 1993.

[48] E.I. Jury, *Theory and Application of the z-transform method*, John Wiley & sons, inc., New York, 1964.

[49] M.H.H. Höfelt, "On the stability of a 1-bit-quantized feedback system," in *Proc. of the Int. Conf. on Acoustics, Speech and Signal Processing (ICASSP)*, Washington, 1979, pp. 844–848.

[50] Richard Schreier, Montgomery V. Goodson, and Bo Zhang, "An algorithm for computing convex positively invariant sets for delta-sigma modulators," *IEEE Trans. on Circuits and Systems I: Fundamental Theory and Applications*, vol. 44, no. 1, pp. 38–44, Jan. 1997.

[51] R.T. Baird and T.S. Fiez, "Stability analysis of high-order delta-sigma modulation for ADC's," *IEEE Trans. on Circuits and Systems-II*, vol. 41, no. 1, pp. 59–61, Jan. 1994.

[52] Kirk C.-H. Chao, Shujaat Nadeem, Wai L. Lee, and Charles G. Sodini, "A higher order topology for interpolative modulators for oversampling A/D converters," *IEEE Trans. on Circuits and Systems*, vol. 37, no. 3, pp. 309–318, Mar. 1990.

[53] Richard Schreier and Martin Snelgrove, "Stability in a general $\Sigma\Delta$ modulator," in *Proc. of the Int. Conference on Acoustics, Speech and Signal Processing (ICASSP)*, Toronto, Canada, May 1991, pp. 1769–1772.

[54] J.A.E.P. van Engelen and B.E. Sarroukh, "Phase uncertainty of a sampled quantizer," in *Proc. of the Workshop on Circuits, Systems and Signal Processing (CSSP)*, Mierlo (NL), 1997, pp. 171–176.

[55] Paul R. Halmos, *Measure Theory*, p. 69, Graduate Text in Mathematics (GTM). Springer Verlag, 1974.

[56] Harold M. Stark, *An Introduction to Number Theory*, pp. 21–22, Markham Publishing Company, Chicago, 1970.

[57] J.A.E.P. van Engelen and R.J. van de Plassche, "New stability criteria for the design of low-pass sigma-delta modulators," in *Proc. of the Int. Symp. on Low Power Electronics and Design (ISLPED)*, Monterey, 1997, pp. 114–118.

[58] L. Longo and B.-R. Horng, "A 15b 30kHz bandpass sigma-delta modulator," in *IEEE Int. Solid-State Circuits Conf. (ISSCC) digest of technical papers*, Feb. 1993, pp. 226–227.

[59] E.J. van de Zwan, "A 2.3mW CMOS ΣΔ modulator for audio applications," in *IEEE Int. Solid-State Circuits Conf. (ISSCC) digest of technical papers*, San Fransisco, 1997, pp. 220–221.

[60] Omid Shoaei and W. Martin Snelgrove, "Design and implementation of a tunable 40MHz-70MHz Gm-C bandpass ΣΔ modulator," *IEEE Trans. on Circuits and Systems II: analog and digital signal processing*, vol. 44, no. 7, pp. 521–530, July 1997.

[61] Erwin Kreyszig, *Advanced Engineering Mathematics*, pp. 860–881, John Wiley & sons, Inc., New York, sixth edition, 1988, ISBN 0-471-62787-9.

[62] Frédéric Gourgue and Maurice Bellanger, "A bandpass subsampled delta-sigma modulator for narrowband cellular mobile communications," in *Proc. of the IEEE. Int. Symp. on Circuits and Systems (ISCAS)*, London, 1994, pp. 353–356.

[63] J. van Engelen and R. van de Plassche, "Stability and design of continuous-time bandpass sigma delta modulators," in *Proc. of the Int. Symp. on Circuits and Systems (ISCAS)*, Orlando (FL), May 1999, vol. II, pp. 355–359.

[64] Altera Corporation, San Jose (CA), *Altera Data Book*, 1995.

[65] J. van Engelen and R. van de Plassche, "A combined LC/gmC 4th order continuous-time bandpass ΣΔ modulator," in *Proc. of the European Solid-State Circuits Conference (ESSCIRC)*, The Hague (NL), 1998, pp. 156–159.

[66] K. Kianush and C. Vaucher, "A global car radio ic with inaudible signal quality checks," in *IEEE Int. Solid-State Circuits Conf. (ISSCC) digest of technical papers*, San Fransisco (CA), Feb. 1998, pp. 130–131.

[67] H.-J. Dressler, "Interpolative bandpass a/d conversion - experimental results," *Electronics Letters*, vol. 26, no. 20, pp. 1652–1653, Sept. 1990.

[68] Waterloo Maple Inc., Waterloo, Canada, *Maple V Handbook - Release 4*, 1996.

[69] J. van Engelen, R. van de Plassche, E. Stikvoort, and A. Venes, "A 6th order continuous time bandpass sigma delta modulator for digital radio IF," in *Int. Solid-State Circuits Conf.*, San Fransisco, 1999, pp. 56–57.

[70] A. van Bezooijen, N. Ramalho, and J.O. Voorman, "Balanced integrator filters at video frequencies," in *Integrated Continuous-Time Filters*, Y.P. Tsividis and J.O. Voorman, Eds., pp. 129–132. IEEE Press, New York, 1993, ISBN 0-7803-0425-X, (reprint from Digest ESSCIRC'91, pp.1-4, 1991).

[71] Augusto Marques, *High Speed CMOS Data Converters*, Ph.D. thesis, Katholieke Universiteit Leuven, Jan. 1999, ISBN 90-5682-166-0.

[72] G. Tröster, H.-J. Dreßler, H.-J. Golberg, W. Schardein, E. Zocher, A. Wedel, K. Schoppe, and J. Arndt, "An interpolative bandpass converter on a 1.2μm BiCMOS analog/digital array," *IEEE Journal of Solid-State Circuits*, vol. 28, no. 4, pp. 471–477, Apr. 1993.

[73] Stephen A. Jantzi, W. Martin Snelgrove, and Paul F. Ferguson, "A fourth-order bandpass sigma-delta modulator," *IEEE Journal of Solid-State Circuits*, vol. 28, no. 3, pp. 282–291, Mar. 1993.

[74] F.W. Singor and M. Snelgrove, "10.7MHz bandpass sigma-delta A/D modulators," in *Proc. of the IEEE 1994 Custom Integrated Circuits Conference (CICC)*, San Diego, CA, May 1994, pp. 163–166.

[75] Bang-Sup Song, "A 4th-order bandpass $\Delta\Sigma$ modulator with reduced number of opamps," in *IEEE Int. Solid-State Circuits Conf. (ISSCC) digest of technical papers*, San Fransisco (CA), Feb. 1995, pp. 204–205.

[76] Theodore Varelas, Seyfi S. Bazarjani, and W. Martin Snelgrove, "A bipolar sampled-data bandpass delta-sigma a/d modulator," in *Proc. of the IEEE Custom Integrated Circuits Conf. (CICC)*, San Diego (CA), May 1996, pp. 205–208.

[77] Orhan Norman, "A band-pass $\Sigma\Delta$ modulator for ultrasound imaging at 160MHz clock rate," in *IEEE Int. Solid-State Circuits Conf. (ISSCC) digest of technical papers*, San Fransisco (CA), Feb. 1996, pp. 236–237.

[78] A. Hairapetian, "A 81MHz IF receiver in CMOS," *IEEE Journal of Solid-State Circuits*, vol. 31, no. 12, pp. 1981–1986, Dec. 1996.

[79] A.K. Ong and B.A. Wooley, "A two-path bandpass $\Sigma\Delta$ modulator for digital IF extraction at 20MHz," in *IEEE Int. Solid-State Circuits Conf. (ISSCC) digest of technical papers*, San Fransisco, 1997, pp. 212–213.

[80] Gopal Raghavan, Joseph F. Jensen, Robert H. Walden, and William P. Posey, "A bandpass $\Sigma\Delta$ modulator with 92dB SNR and center frequency continuously programmable from 0 to 70MHz," in *IEEE Int. Solid-State Circuits Conf. (ISSCC) digest of technical papers*, San Fransisco (CA), Feb. 1997, pp. 214–215.

[81] Stephen A. Jantzi, Kenneth W. Martin, and Adel S. Sedra, "Quadrature bandpass $\Delta\Sigma$ modulation for digital radio," *IEEE Journal of Solid-State Circuits*, vol. 32, no. 12, pp. 1935–1950, Dec. 1997.

[82] Weinan Gao and W. Martin Snelgrove, "A 950-MHz IF second-order integrated LC bandpass delta-sigma modulator," *IEEE Journal of Solid-State Circuits*, vol. 33, no. 5, pp. 723–732, May 1998.

[83] H. Tao and J. Khoury, "A 100MHz IF, 400MSample/s CMOS direct-conversion bandpass $\Sigma\Delta$ modulator," in *IEEE Int. Solid-State Circuits Conf. (ISSCC) digest of technical papers*, San Fransisco (CA), Feb. 1999, pp. 60–61.

[84] A. Namdar and B. Leung, "A 400MHz 12b 18mW IF digitizer with mixer inside a $\Sigma\Delta$ modulator loop," in *IEEE Int. Solid-State Circuits Conf. (ISSCC) digest of technical papers*, San Fransisco (CA), Feb. 1999, pp. 62–63.

[85] Loai Louis, John Abcarius, and Gordon W. Roberts, "An eigth-order bandpass $\Delta\Sigma$ modulator for A/D conversion in digital radio," *IEEE Journal of Solid-State Circuits*, vol. 34, no. 4, pp. 423–431, Apr. 1999.

[86] Georgi P. Petkov and Anthony C. Davies, "Constraints on constant-input oscillations of a bandpass sigma-delta modulator structure," *Int. Journal of Circuit Theory and Applications*, vol. 25, no. 5, pp. 393–405, Aug.-Sep. 1997.

[87] William H. Press, Brian P. Flannery, Saul A. Teukolsky, and William T. Vetterling, *Numerical Recipes in Pascal; The Art of Scientific Computing*, Cambridge University Press, Cambridge / New York / Melbourne, 1989, ISBN 0-521-37516-9.

LIST OF ACRONYMS

A/D	Analog-to-Digital
ADC	Analog-to-Digital Converter
AGC	Automatic Gain Control
AM	Amplitude Modulation
BI	Balanced Integrator
BIBS	Bounded Input Bounded Output
BPF	Bandpass Filter
BW	Bandwidth
cf	characteristic function
CSD	Canonical Signed Digit
CT	Continuous Time
D/A	Digital-to-Analog
DAC	Digital-to-Analog Converter
DC	Direct Current
DCV	Direct Conversion
df	describing function (method)
DNL	Differential Non Linearity
DR	Dynamic Range
DSP	Digital Signal Processing
DT	Discrete Time
ECL	Emitter Coupled Logic
ENOB	Effective Number of Bits
FFT	Fast Fourier Transform
FM	Frequency Modulation
FOM	Figure of Merit
gcd	greatest common divider
gmC	transconductance-capacitor
IC	Integrated Circuit
IF	Intermediate Frequency
IM3	Third order Intermodulation Product
INL	Integral Non Linearity
IP3	Third order Intermodulation Intercept Point
LC	inductor-capacitor
LNA	Low Noise Amplifier
LO	Local Oscillator
MASH	Multi Stage (Noise Shaping)
NRZ	Non Return to Zero
NTF	Noise Transfer Function
OSR	Oversampling Ratio

P.A.	Power Amplifier
PC	Personal Computer
P/S	Parallel to Serial
psd	Power Spectral Density
RF	Radio Frequency
rms	root mean square
RTZ	Return to Zero
SC	Switched Capacitor
SDM	Sigma Delta Modulator
SFDR	Spurious Free Dynamic Range
SNDR	Signal to Noise and Distortion Ratio
SNR	Signal to Noise Ratio
STF	Signal Transfer Function
T/H	Track and Hold
THD	Total Harmonic Distortion
TI	Time Interleaved
VHDL	Verilog Hardware Description Language

LIST OF SYMBOLS

α	quantizer stability model phase uncertainty parameter
γ	frequency normalized to baseband signal bandwidth f_b
δ	Dirac operator
ε	enclosure of a set
λ	quantizer stability model gain parameter
θ	angular frequency
θ_b	angular baseband bandwidth
θ_0	angular tuning frequency
ω_0	tuning frequency
ϕ	phase
Φ	set of phases
a	SDM loop filter zeroes parameter; NTF pole radius
A	signal amplitude
b	SDM loop filter pole radius; NTF zero radius
B	number of bits; resolution of a quantizer in bits
BW	bandwidth
c_g	quantizer noise model parameter
e_q	quantization error signal
E_q	power spectral density (spectrum) of e_q
$E_q(z)$	z-transform of e_q
f	frequency
f_b	baseband signal bandwidth
f_D	bandwidth of frequency modulated distortion components (tones)
f_i	input frequency
f_H	frequency distortion component (tone)
f_L	frequency distortion component (tone)
f_0	tuning frequency
f_s	sample frequency
$F(p)$	continuous time loop filter
$F(z)$	equalization filter (MASH)
G_{DC}	integrator DC gain
$G(z)$	SDM loop filter
$G_{eq}(z)$	equivalent discrete time loop filter
$H(z)$	noise shaper loop filter
i	input signal
$I(z)$	z-transform of input signal
j	complex unit; $j^2 = -1$
J_p	Bessel function
k	integer; discrete time index

k_b	Boltzmanns constant; $1.38 \cdot 10^{-23}$ J/K
K	integer; number of samples used for the phase uncertainty approximation
K_{ssf}	subsample factor
$\mathcal{L}$	Laplace transform operator
$\mathbb{N}$	set of natural numbers
N	integer; order of loop filter
N_i	equivalent input noise power
N_{ADC}	noise contribution of the ADC
N_{DAC}	noise contribution of the DAC
N_{fil}	noise contribution of the filter
N_o	total in-band noise power
N_q	quantization noise density
o	quantizer output signal
$O(z)$	z-transform of quantizer output signal
OSR	oversampling ratio
p	integer; Laplace transform variable
P	power consumption
$P^{\star}$	number of quantization levels
P_i	input power
q	quantizer step size
$\mathbb{Q}$	set of rational numbers
Q	quality factor
$\mathbb{R}$	set of real numbers
$R(p)$	Laplace transform of DAC pulse
S_c	quantizer maximum amplitude correction factor
t	continuous time
T	period (time)
T_s	sample period
$V(.)$	Lyapunov function
x	quantizer input signal
$\mathbf{x}$	state vector
$X(z)$	z-transform of quantizer signal
z	z-transform variable
$\mathbb{Z}$	set of whole numbers
$\mathcal{Z}$	z-transform operator

APPENDIX A

MODULATOR RESPONSE TO INPUT SIGNALS

The nonlinearity of the quantizer inside the loop of a sigma delta modulator (see Fig. A.1) causes the behavior of sigma delta modulators to be strongly input signal dependent. The range of the states of the SDM and the stability properties of the SDM not only depend on the amplitude of the input signal, but also on the frequency of the input signal. This can be easily shown by the following experiments. First, the range of the filter output values of a second order bandpass SDM will be examined as a function of the amplitude of the input sine wave. The loop filter of the SDM is a second order resonator. In the z-domain the loop filter transfer function is given by:

$$G(z) = \frac{-2r\cos(\theta_0)z^{-1} - r^2z^{-2}}{1 - 2r\cos(\theta_0)z^{-1} + r^2z^{-2}} \tag{A.1}$$

The transfer of the filter in discrete time domain can be written as:

$$x[k] = 2r\cos(\theta_0)x[k-1] - r^2x[k-2] + 2r\cos(\theta_0)u[k-1] - r^2u[k-2] \tag{A.2}$$

in which $u[k]$ and $x[k]$ are the input and output of the filter. The parameter r denotes the radius of the poles and determines the "damping" of the resonator; θ_0 represents the tuning frequency of the resonator. For $r \leq 1$, the SDM is generally considered to be stable for all bounded input signals. In [86] constant-input oscillation bounds for the states of this modulator were analyzed. Certain parameter ranges were derived for which the constant-input oscillations were bounded by some area in state space. In the conclusion of this article, the authors stated: "Although these bounds do not apply for time-varying inputs, they may be expected to give some indications of the behavior with such inputs". Analysis of the behavior of the SDM with a time varying input signal is complicated and oscillation bounds for the states are hard to derive. However, an indication of these bounds can be easily obtained by simulation. Plotting the absolute maximum value of the loop filter output as a function of the input sine wave amplitude

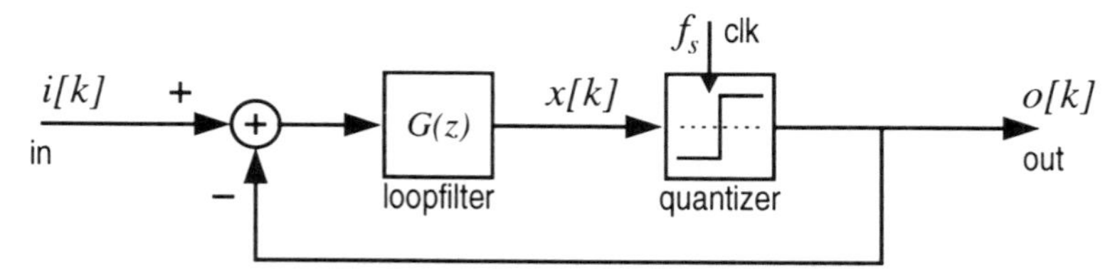

Figure A.1: *A single loop one-bit discrete time sigma delta modulator.*

for several input frequencies can show whether the behavior is frequency dependent or not. As an example, Fig. A.2 shows such a plot for an SDM with a pole radius $r = 1$ and a tuning frequency $\theta_0 = 0.6$ rad. The amplitude is shown relative to the quantizer step size q (the quantizer output has a value of $\pm q/2$). For a DC input value ($\theta = 0$) and

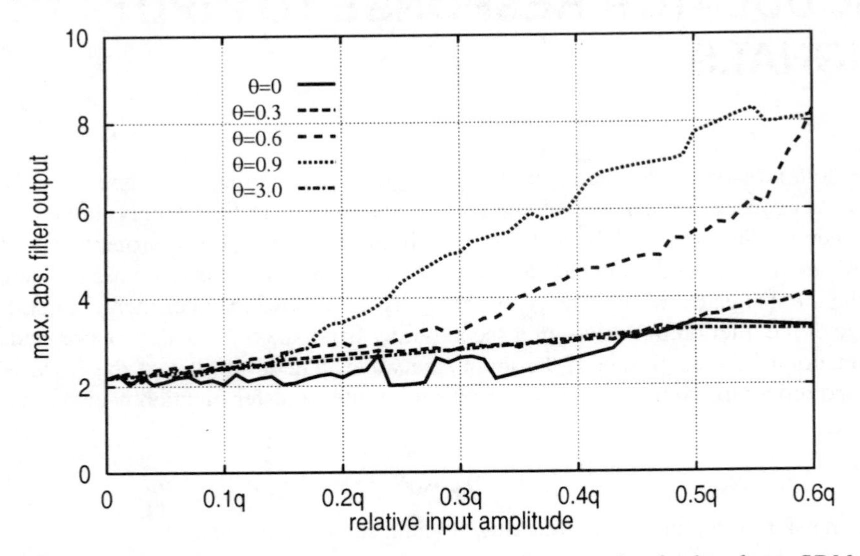

Figure A.2: *Maximum absolute filter output of a second order bandpass SDM as a function of the input amplitude, for several input frequencies.*

other frequencies not near to the tuning frequency ($\theta = 0.3$ and $\theta = 3.0$), the maximum absolute filter output value does not vary significantly with the input amplitude. For an input frequency equal to the tuning frequency of the SDM, the maximum absolute output value of the filter increases considerably when the amplitude approaches $q/2$. In the case that the input frequency ($\theta = 0.9$) is slightly higher than the tuning frequency, the loop filter output shows an even larger increase. This experiment shows that the dynamic behavior of an SDM strongly depends frequency of the input signal. The expectation that constant-input analysis may give an indication of the behavior for time-varying inputs is therefore unfounded.

A second experiment shows the signal dependent stability of a sigma delta modulator. The maximum stable input amplitude for a sixth order bandpass SDM will be determined as a function of the input frequency. The loop filter of the bandpass SDM is given by:

$$G(z) = \frac{(1-0.6z^{-2})^3}{(1-z^{-2})^3} - 1 \tag{A.3}$$

The loop filter is tuned at $f_s/4$. Figure A.3 shows the maximum stable input amplitude as a function of the input frequency. The modulator is considered to be stable when the absolute value of the output of the loop filter does not exceed a certain threshold

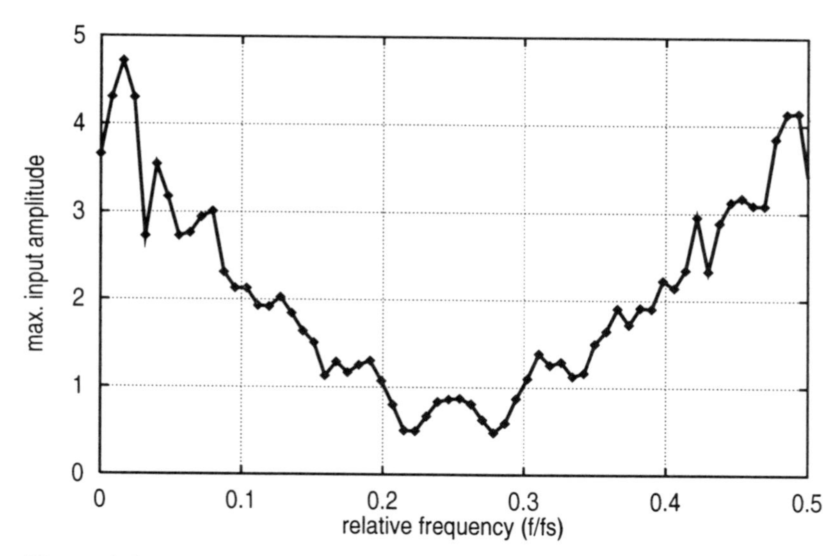

Figure A.3: *Maximum stable input amplitude of a sixth order bandpass SDM as a function of the input frequency.*

value during the simulation. When the loop filter output exceeds the threshold, the behavior of the SDM is assumed to be determined by a large-signal limit cycle (see chapter 5) and the modulator is considered to be unstable. In the case of Fig. A.3, the threshold value was chosen 1000 times the quantizer output value $q/2$ and a simulation of 10^6 points was performed. As can be seen in Fig. A.3, the maximum allowed input amplitude for stability strongly depends on the frequency of the input signal. For signals having a frequency near the tuning frequency $f_s/4$, the maximum allowed amplitude is about a factor ten smaller than for signals with an out-of-band frequency. The worst-case maximum signal amplitude occurs for in-band signals. Experiments determining the stability properties of an SDM should therefore use in-band signals.

The two experiments above support the intuitive notion that input signals for which the loop filter of the sigma delta modulator exhibits a high gain have more impact on the dynamic behavior of the SDM than input signals which are not within the pass band of the loop filter. In the case the frequency of the input signal falls within the pass band of the filter, the quantizer input signal is strongly affected by the applied input. Consequently, the applied input signal also strongly affects the output of the quantizer and the behavior of the unexcited SDM (e.g. an idle pattern). When the frequency of the input signal falls outside the pass band of the loop filter, a much larger amplitude is required to have the same amount of effect on the quantizer input signal.

APPENDIX B

ROOT LOCUS SEARCH METHOD

The stability of the SDMs is investigated by analyzing the root loci of a linearized model (see chapter 5). The linearized model contains two model parameters: a gain λ and a phase uncertainty parameter α (see Fig. B.1). The root locus is the plot of the

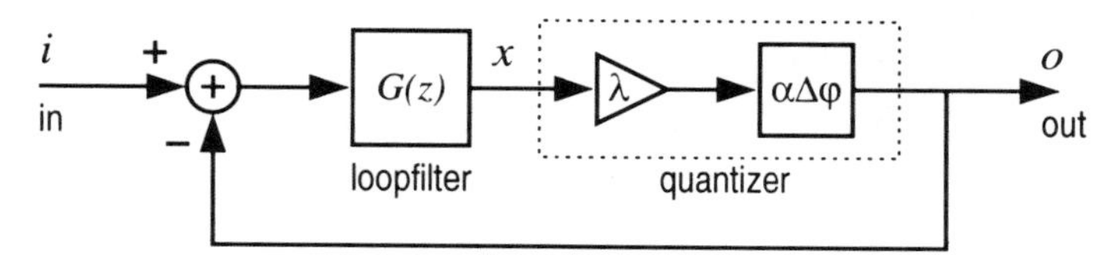

Figure B.1: Stability Model of a sigma delta modulator.

poles of the (linearized) closed-loop system of Fig. B.1 in the complex z-plane as a function of the model parameters. The poles of the closed loop system are equal to the roots of the 'stability equation':

$$1+\lambda e^{j\alpha\Delta\phi(\theta)}G(z)=0 \quad \text{with:} \quad z=r\cdot e^{j\theta} \tag{B.1}$$

Because of the term $e^{\alpha\Delta\phi(\theta)}$, the roots of this equation cannot be solved analytically for $\alpha \neq 0$. Therefore, a search method will used to scan the complex z-plane for solutions. For this search method, the following assumptions are made:

- The root locus will be solved for a certain value of the phase uncertainty parameter $\alpha \in [-1,1]$ and all allowed values of the gain parameter $\lambda \in [0,\infty\rangle$. The basic thought behind this choice is that instability will give rise to a change in the amplitude of the signal inside the loop, and hence a change of λ. The phase parameter α is considered to be unaffected.

- For a constant value of α, the location of the roots of eq. (B.1) will be a continuous function of λ. In other words, for a constant value of α, the trajectories of the poles of the closed-loop system in the z-plane as a function of λ do not contain any discontinuities for $0<\lambda<\infty$. Note that this assumption does not prohibit so-called break- and merge-points in the root loci.

The concept behind the search method is that when $z=z^\star$ is a solution of the stability equation for some λ and α, the corresponding value for λ should be real-valued. For $z=z^\star$ and a certain α, the value for λ can be calculated using eq. (B.1):

$$\lambda = \frac{-1}{e^{j\alpha\Delta\phi(\theta^\star)}G(z^\star)} \tag{B.2}$$

In the case that $z^\star$ is a solution of the stability equation, the imaginary part of the left hand side of eq. (B.2) should be zero:

$$\mathrm{Im}(\lambda) = \mathrm{Im}\left\{\frac{-1}{e^{j\alpha\Delta\phi(\theta^\star)}G(z^\star)}\right\} = 0 \tag{B.3}$$

A crude but effective method to find a solution $z^\star$ satisfying this requirement is depicted in Fig. B.2. The imaginary part of λ is evaluated using eq. (B.2) for successive points on a scan line in the complex plane. When the imaginary part of λ changes sign

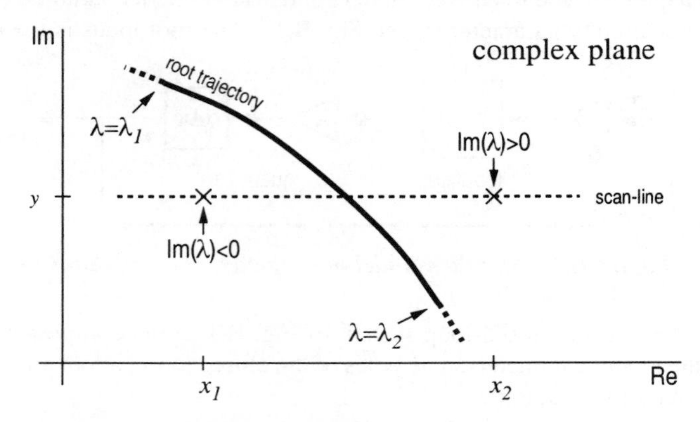

Figure B.2: *Finding a root in the complex plane.*

between two evaluated points on the scan-line, two solutions are possible: (i) a singularity for the imaginary part of λ occurs for an intermediate point on the scan line, or (ii) the imaginary part of λ is zero for an intermediate point on the scan line. The latter solution corresponds to a solution of the stability equation. The intermediate point for which the imaginary part of λ is zero, i.e. λ is real, corresponds to an intersection point of the root trajectory and the scan line (see Fig. B.2). At this point, finding a complex-valued solution to the stability equation (B.1) has been reduced to a common real-valued root finding problem:

$$f(x) = \mathrm{Im}\left\{\frac{-1}{e^{j\alpha\Delta\phi(x+jy)}G(x+jy)}\right\} = 0 \quad \text{for:} \quad x_1 < x < x_2 \tag{B.4}$$

This problem can be solved by well known methods such as the bisection method [87]. When a sufficiently accurate value $x = x^\star$ has been found, $z^\star = x^\star + jy$ is a point on the root locus. The corresponding (real) value of λ can be calculated by eq. (B.2).

By structured scanning of a selected area of the complex z-plane, sufficient points on the root locus can be found and plotted. For the root loci in this book, a square area of a selected size was scanned using densely placed horizontal and vertical scan lines. Note that the positioning of the scan line is not limited to a horizontal or vertical orientation. In particular, a similar scan can be performed along the unit circle, thus providing an accurate estimate of the unit circle crossings of the root locus.

An example of the root locus search method is shown in Fig. B.3. Here the root locus of a third order SDM is shown for $\alpha = -1$. The transfer of the loop filter of the modulator is given by:

$$G(z) = \frac{(1-0.5z^{-1})^3}{(1-z^{-1})^3} - 1 \tag{B.5}$$

For this example an area of two by two centered around the origin of the complex plane was scanned using ten horizontal and ten vertical scan lines. The points of the

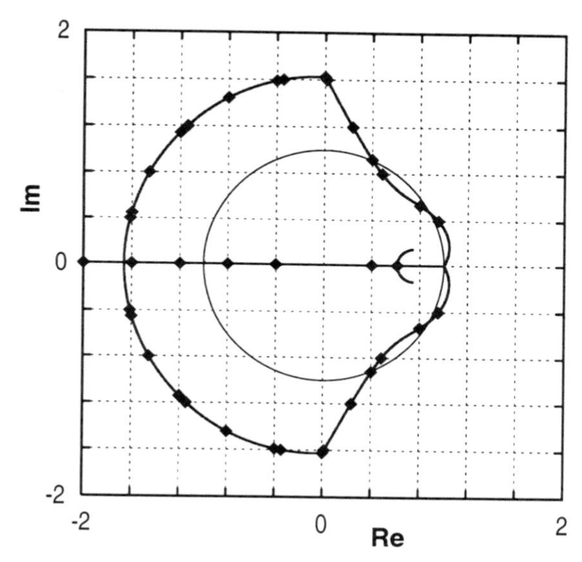

Figure B.3: Example of the root locus search method (3rd order lowpass SDM).

root locus found in this way are indicated by large dots. A more dense scan (400 horizontal and 400 vertical scan lines) was also performed to show the overall shape of the root locus.

APPENDIX C

ALGORITHM FOR FINDING A STABILITY BOUNDARY

The stability boundary for a parameter a of a lowpass SDM loop filter can be calculated by determining the value for a for which the root locus with the worst-case phase uncertainty is tangent to the unit circle (see Chapter 5). The intersection points of the root locus with the unit circle are determined by the stability equation

$$1 + \lambda e^{j\alpha\Delta\phi(\theta)} G(z) = 0 \quad \text{with:} \quad z = e^{j\theta} \tag{C.1}$$

A point $z = e^{j\theta}$ on the unit circle is part of the root locus when eq. (C.1) is satisfied for a real-valued λ (see App. B). Rewriting eq. (C.1) and substituting the worst-case phase uncertainty value $\alpha = -1$ gives

$$\lambda = \frac{-1}{e^{-j\Delta\phi(\theta)} G(e^{j\theta})} \tag{C.2}$$

Analysis of the root loci of lowpass modulators shows that when a is large enough, the worst-case root locus ($\alpha = -1$) has at least two intersection points with the unit circle between $\theta = \pi/2$ and $\theta = \pi/6$ (see Chapter 5). This means that the imaginary part of (C.2) intersects the x-axis twice as shown in Fig. C.1 (left). By decreasing the value of the loop filter parameter a, the two intersection points of the root locus with the unit circle will move towards each other until for some $a = a^{\star}$, there is only one intersection point with the unit circle. This means the imaginary part of λ has exactly one crossing with the x-axis (see Fig. C.1(right)). Consequently, the derivative of the imaginary part of λ as a function of θ will be zero. The value for a for which the worst-case root locus is tangent to the unit circle can be calculated by solving:

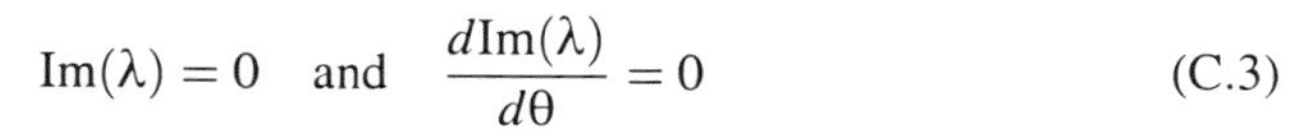

$$\text{Im}(\lambda) = 0 \quad \text{and} \quad \frac{d\text{Im}(\lambda)}{d\theta} = 0 \tag{C.3}$$

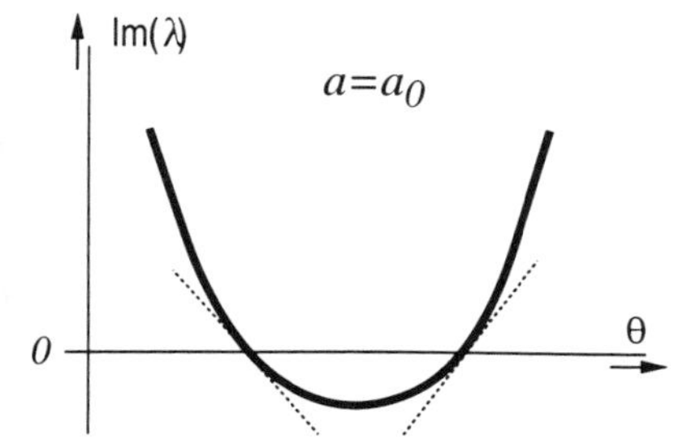

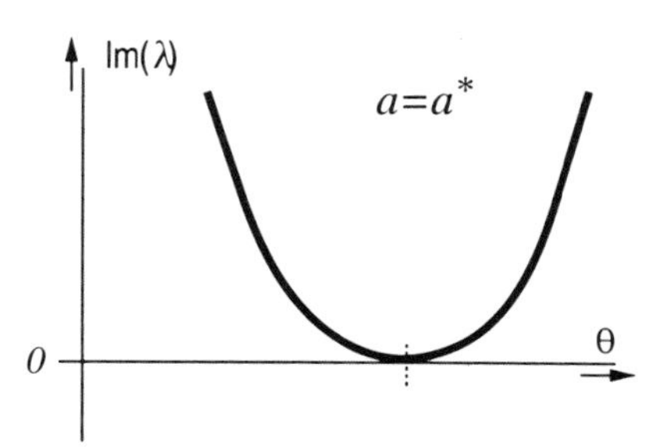

Figure C.1: *Finding a single root locus intersection point.*

in which λ is given by (C.2). This set of two equations can be solved numerically using a nested Newton-Raphson algorithm [87]. Note that nesting of Newton-Raphson algorithms is generally not recommended for solving a set of equations such as (C.3)[1]. In this particular case, the nested algorithm will be convergent towards an accurate solution when the required initial guesses are chosen wisely. The resulting algorithm is as follows:

1. select a initial guess for a such that the worst-case root locus has two intersection points with the unit circle.

2. select an initial guess for θ for which the worst-case root locus will be tangent to the unit circle.

3. Calculate the value for θ for which $d\text{Im}(\lambda)/d\theta = 0$ using Newton-Raphson.

4. Calculate the actual value of $\text{Im}(\lambda)$. and determine the required change in the loop filter parameter a using the Newton Raphson method.

5. If the change in a is larger than the required accuracy in the final solution of a, another iteration step is required: go back to 2; otherwise the desired value for which the root locus is tangent to the unit circle has been found with the required accuracy.

[1] The Newton-Raphson algorithm for solving $f(x) = 0$ uses the derivative $df(x_i)/dx_i$ to calculate a new iterate x_{i+1} which will be closer to the solution $x^\star$ for which $f(x^\star) = 0$. In the case that $df(x_i)/dx_i$ also goes to zero for $x_i \to x^\star$, convergence of the Newton-Raphson algorithm is not guaranteed.

APPENDIX D

EXAMPLE VHDL DESCRIPTION OF A DIGITAL SDM

Using the test set-up described in section 7.1, an all-digital sigma delta modulator can be simulated in real time. The behavior of digital set-up can be programmed in VHDL. In this appendix an example VHDL description is shown of a (randomly chosen) fourth order bandpass one-bit sigma delta modulator. The loop filter of the modulator is described by:

$$G(z) = \frac{(1-az^{-1})^4}{(1-2b\cos(\theta_0)z^{-1}-b^2z^{-2})^2} - 1 \tag{D.1}$$

with $a = 0.25$, $b = 1$ and $\theta_0 = 0.3352\pi = 1.0521$ rad. Writing out the coefficients gives:

$$G(z) = \frac{0.97965z^{-1} - 2.60475z^{-2} + 1.91715z^{-3} - 0.99609z^{-4}}{1 - 1.97965z^{-1} + 2.97975z^{-2} - 1.97965z^{-3} + z^{-4}} \tag{D.2}$$

The four digit Canonical Signed Digit (see sec. 7.1) approximation of the coefficients equals:

$$0.97965 \approx 1.00000\mathrm{T0T} = 1 - \frac{1}{64} - \frac{1}{256} = 0.98046 \tag{D.3}$$

$$2.60475 \approx 10.10100\mathrm{T} = 2 + \frac{1}{2} + \frac{1}{8} - \frac{1}{64} = 2.60937 \tag{D.4}$$

$$1.91715 \approx 10.000\mathrm{T0T0T} = 2 - \frac{1}{16} - \frac{1}{64} - \frac{1}{256} = 1.91797 \tag{D.5}$$

$$0.99609 \approx 1.0000000\mathrm{T} = 1 - \frac{1}{256} = 0.99609 \tag{D.6}$$

$$1.97965 \approx 10.00000\mathrm{T0T} = 2 - \frac{1}{64} - \frac{1}{256} = 1.98047 \tag{D.7}$$

$$2.97975 \approx 11.00000\mathrm{T0T} = 2 + 1 - \frac{1}{64} - \frac{1}{256} = 2.98047 \tag{D.8}$$

These coefficients can be recognized in the `filter` section of the VHDL description in Fig. D.1. The modulator is described by an **`entity`** definition and a corresponding **`architecture`**. Apart from the `filter` section, the **`architecture`** contains some input and output register descriptions, a `quantizer` section and input and output control (`modinp` and `whichoutput`). In addition to an signal input, a signal output and a clock input port, the SDM also has three control inputs. The control line `WO` determines which signal is send to the DAC. The DAC is connected to either the quantizer output or the loop filter output signal. The control line `ZI` allows the input to be exactly zero. This is useful for examining idle patterns, as the input ADC may exhibit noisy behavior of the LSB bit. The final control line `SI` is used to scale the input. By making this control line low, the input is scaled down by 5 bits corresponding to an attenuation of approximately 30dB.

```
-- -------------------- VHDL DESCRIPTION STARTS HERE -------------------- --
ENTITY bp2 IS
   PORT (clk:   IN BIT;                                    -- clock:  1 bit
         inp:   IN INTEGER RANGE -2048 TO 2047;            -- filter input:  12 bit
         FOUTP: OUT INTEGER RANGE -2048 TO 2047;           -- filter output:  12 bit
         WO: IN BIT;                                       -- which output
         SI: IN BIT;                                       -- scale input ( 5 bit = 30 dB)
         ZI: IN BIT                                        -- zero input
         );
END bp2;

ARCHITECTURE structure OF bp2 IS

SUBTYPE ext_sig IS INTEGER RANGE -2048 TO 2047;            -- 12 bit external
SUBTYPE int_sig IS INTEGER RANGE -134217728 to 134217727;  -- 28 bit internal
SIGNAL inpff,foutpff:  ext_sig;                            -- external states:  flipflops
SIGNAL x1,x2,x3,x4,filout,x1tmp:  int_sig;                 -- internal states:  flipflops
SIGNAL newx1,newx2,newx3,newx4:  int_sig;                  -- new internal states
SIGNAL qout:  INTEGER RANGE -1 TO 1;                       -- quantizer output
SIGNAL filinp:  INTEGER RANGE -8388608 TO 8388607;         -- filter input 23 bit
SIGNAL inp2:  INTEGER RANGE -4194304 TO 4194303;           -- scaled SDM input 22 bit
SIGNAL ziff,siff,woff:  BIT;                               -- control signal flipflops

BEGIN
statesinoutregs:  PROCESS(clk)
   BEGIN
      IF (clk'EVENT) AND (clk='1') THEN                    -- clock new states
      x1<=newx1;
      x2<=newx2;
      x3<=newx3;
      x4<=newx4;
      END IF;
      IF (clk'EVENT) AND (clk='0') THEN                    -- clock new inputs
      inpff<=inp;
      ziff<=zi;
      siff<=si;
      woff<=wo;
      foutp<=foutpff;
      END IF;
   END PROCESS;
```

Figure D.1: *VHDL SDM Description (part 1).*

```
filter:  PROCESS(x1,x2,x3,x4,filinp)                             -- calculate new states
   BEGIN
      x1tmp<=(2*x1 - x1/64 -x1/256);
      filout<=x1;
      newx1<=x2 + (filinp - filinp/64 - filinp/256) + x1tmp;
      newx2<=x3 - (2*filinp + filinp/2 + filinp/8 - filinp/64) - (x1tmp+x1);
      newx3<=x4 + (2*filinp - filinp/16 -filinp/64 - filinp/256) + x1tmp;
      newx4<=0 - (filinp - filinp/256) - x1;
   END PROCESS;

quantizer:  PROCESS(filout)                                      -- quantizer operation
   BEGIN
      IF (filout>=0) THEN qout<=1; ELSE qout<=-1; END IF;
   END PROCESS;

modinp:  PROCESS(inpff,ziff,siff)
   BEGIN
      IF (ziff='1') THEN inp2<=0;                                -- zero input
      ELSIF (siff='1') THEN inp2<=inpff*2048;                    -- full scale input
      ELSE inp2<=inpff*64;                                       -- scale input 5 bits (30dB)
      END IF;
   END PROCESS;

feedback:  PROCESS(inp2,qout)
   BEGIN
      filinp<=inp2-qout*4194304;                                 -- scale quantizer feedback level
   END PROCESS;

whichoutput:  PROCESS(filout,woff,qout)
   BEGIN
      IF (woff='1') THEN foutpff<=filout/65536;                  -- select filter output (scaled)
      ELSE foutpff<=qout*1024; END IF;                           -- select quantizer output (scaled)
   END PROCESS;
END structure;
-- --------------------- VHDL DESCRIPTION ENDS HERE --------------------- --
```

Figure D.1: VHDL SDM Description (continued).

ABOUT THE AUTHORS

Jurgen van Engelen was born in Rijen, the Netherlands in April 1971. In 1995, he received the M.S. degree in Electrical Engineering from the Eindhoven University of Technology, Eindhoven, the Netherlands on the subject of analog cellular neural networks. In May 1999, he received his Ph.D. degree from the same university for his Ph.D. thesis: "Stability Analysis and Design of Bandpass Sigma Delta Modulators." In July 1999, he joined Broadcom Corp., Irvine, CA as a staff scientist in the Analog and RF Microelectronics Group. His research interests comprise stability issues and design of high-speed analog electronics for advanced VLSI communication circuits. He has published and presented papers at several international conferences.

Rudy van de Plassche was born in IJzendijke, The Netherlands, on September 24, 1941. In 1964, he graduated from Delft University of Technology. In 1989, he obtained his Ph.D. from the same univiersity. The title of his thesis was: "High-Speed and High-Resolution A/D and D/A Conversion." In 1964, he joined Philips Research Laboratories in Eindhoven where he was involved in circuit design for analog integrated circuits. His research interests are in the field of operational amplifiers, instrumentation amplifiers, analog multipliers, integrated reference sources, high-speed and high-resolution analog-to-digital and digital-to-analog converters, sample-and-hold amplifiers, and lately digital signal processing circuits and systems. In 1983, he transferred to Philips Research Laboratories, Sunnyvale, CA, where he became Group Manager of the Advanced Design group. In 1986, he returned to Philips Research in Eindhoven, where he became a member of the "Integrated Tranceivers" group, involved in the design of digital radio systems and digital signal processing. Starting September 1, 1989, he was appointed as a part-time professor at the Eindhoven University of Technology in the department of "Signal Processing and Electronic Systems," responsible for "Telecommunication circuits." In May 1998, he became a full-time professor in the "Electronic Circuit Design Group." Since July 1999, he is a Technical Director at Broadcom Netherlands B.V. in Bunnik. He holds 65 U.S. patents, published more than 46 papers in IEEE journals and presented papers at several international conferences. In 1989, Dr. van de Plassche was elected Fellow of the Institute of Electrical and Electronic Engineers. He received the "Veder prijs" of the "Stichting Wetenschappelijk Radiofonds Veder" for the year 1988. In 1996, he received the "Solid State Circuits Award." Since 1983, he has been a member of the Technical Program Committee of the International Solid-State Circuits Conference. In this committee he served as a chairman of the analog subcommittee. Today, he is chairman of the European ISSCC Program Committee and a member of the technical program committee of the European Solid-State Circuits Conference.

INDEX